ADVANCED DESIGN CONCEPTS for ENGINEERS

HOW TO ORDER THIS BOOK

BY PHONE: 800-233-9936 or 717-291-5609, 8AM–5PM Eastern Time

BY FAX: 717-295-4538

BY MAIL: Order Department
Technomic Publishing Company, Inc.
851 New Holland Avenue, Box 3535
Lancaster, PA 17604, U.S.A.

BY CREDIT CARD: American Express, VISA, MasterCard

BY WWW SITE: http://www.techpub.com

ADVANCED DESIGN CONCEPTS for ENGINEERS

B. S. Dhillon, Ph.D., P.E.
Department of Mechanical Engineering
University of Ottawa

LANCASTER · BASEL

Advanced Design Concepts for Engineers
a **TECHNOMIC**® publication

Published in the Western Hemisphere by
Technomic Publishing Company, Inc.
851 New Holland Avenue, Box 3535
Lancaster, Pennsylvania 17604 U.S.A.

Distributed in the Rest of the World by
Technomic Publishing AG
Missionsstrasse 44
CH-4055 Basel, Switzerland

Printed in the United States of America
10 9 8 7 6 5 4 3 2 1

Main entry under title:
Advanced Design Concepts for Engineers

A Technomic Publishing Company book
Bibliography: p.
Includes index p. 237

Library of Congress Catalog Card No. 98-60043
ISBN No. 1-56676-626-5

To the memory of my late great grandfather's brother, Gurdit S. Dhillon

Table of Contents

Preface

IN the discipline of engineering, the term *Design* may convey different meanings to different people. For example, to some people a designer is a person who makes use of the drawing board to draft the detail of an engineering part. Others may think of design as the creation of a sophisticated system, for example, an aircraft. For our purpose, the term *design* simply means the design of items of an engineering nature: structures, products, parts, and so on.

People have been designing engineering-related objects and structures for thousands of years. Egyptian pyramids and the Great Wall of China are two prime examples of ancient civil engineering structures. A treatise on architecture by Vitruvis (a Roman architect) was written in 30 B.C. Since the days of the Industrial Revolution, a vast number of publications relating to engineering design has appeared.

Today's products under design have not only become more complex and sophisticated, but also have to satisfy consumers from various different perspectives in order to survive in the competitive market environment. To meet such a challenge, over the years many new design-related concepts and aids have been developed. These include concurrent engineering, value engineering, configuration management, reliability, maintainability, safety engineering, human engineering, reverse engineering, reengineering, total quality management, life cycle costing, computer-aided design, and the information superhighway. Professionals working in a design environment find it difficult to obtain information on these areas in a single text. Thus, this book is an attempt to present these modern concepts and aids in a single volume after filtering through relevant published literature.

The book is intended for readers such as engineering undergraduate and graduate students, practicing engineers, college- and university-level teachers, design managers, and so forth. In general, emphasis is on the structure of concepts rather than on mathematical rigor and minute detail. The sources of most of the material presented are given in references, in case the reader wishes to delve deeper into a specific area. In addition, the book contains many examples along with their corresponding solutions at appropriate places to facilitate easy understanding of the material presented.

The book itself is self-contained; for example, Chapters 1 and 2 present introductory material on design and design-related mathematics, respectively; the volume is composed of 11 chapters. Chapter 1 briefly discusses various introductory aspects of engineering design, and Chapter 2 reviews basic design-related mathematics useful in understanding subsequent chapters. Chapter 3 discusses three modern areas related to design: concurrent engineering, reverse engineering, and reengineering. Chapter 4 is devoted to design reliability, and some of the topics covered are reasons for considering reliability in product design, reliability in system life cycle, common reliability networks, design evaluation concepts, and reliability and maintainability standard documents. The next two chapters, Chapters 5 and 6, are concerned with maintainability engineering and safety engineering, respectively. Basically, Chapter 5 provides important aspects of maintainability related to engineering design, and Chapter 6 covers topics such as hazard classifications, safety in system life cycle, safety analysis methods, safety costing, and product liability. Chapter 7 is totally devoted to human factors in design. Some of the topics covered are human factor considerations in product design, human behavioral expectancies, human sensory capabilities and body measurements, and formulas for design specialists. Chapter 8 discusses the subject of total quality management (TQM) by describing specifically the topics such as TQM principles and elements, quality in the design phase, and TQM tools. Two important subjects related to engineering design are described in Chapter 9, i.e., value engineering and configuration management. Chapter 10 describes the modern concept of life cycle costing. Some of the subjects described are steps for life cycle cost analysis, costs generated over the product life cycle, and life cycle cost estimation models. Chapter 11 covers two topics: computer-aided design and the information superhighway. Each of these two subjects is discussed in a considerable depth.

The author wishes to thank the editorial department members at Technomic Publishing Co. Inc. for their penetrating input. The author is indebted to his colleagues, students, and others for their interest in this project. I am grateful to Josee Rocheleau of the Mechanical Engineering Department, University of Ottawa, for typing the first draft of this book. I also greatly appreciate the patience and intermittent disturbances of my children, Jasmine and Mark, resulting in many coffee breaks that helped to produce new schools of thought! Last, but not least, I thank my wife Rosy for typing some portions of this project and for her help in proofreading. During the preparation of the manuscript, her patience and tolerance were greatly appreciated.

CHAPTER 1

Introduction to Design

1.1 INTRODUCTION

THE term *design* may convey different meanings to different people. For example, *Webster's Encylopedic Dictionary* [1] gives several different meanings of the term *design*: to prepare plans or a sketch or model of something to be made, to invent and bring into being, to plan in mind, etc. Specifically, in engineering the term *design* also means different things to different people [2]. For example, to some a designer is a person who employs drawing tools to draw the details of an item, and to others, a design is the creation of a complex product such as computer, aircraft, automobile, or air traffic control system.

The history of engineering design can be traced back to the construction of the first known Egyptian pyramid, Saqqara, in 2650 B.C. It was built by Imhotep, who, thus, was perhaps the first design engineer. Because engineering drawings are considered to be the backbone of modern engineering designs, their history could be traced back to 4000 B.C. when the Chaldean engineer Gudea engraved the plan view of a fortress on a stone tablet [3]; thus, it is the oldest surviving technical drawing. Nevertheless, the written evidence of use of technical drawings only goes back to the Roman period; for example, Vitruvius, a Roman architect, wrote a treatise on architecture in 30 B.C. In a modern sense, the first book on engineering drawings, entitled *Geometrical Drawings*, was written by an American named William Minifie in 1849. Since that time a great many people have contributed to both engineering design and drawings, and this chapter describes various introductory aspects of engineering design.

1.2 REASONS FOR DESIGNING: DESIGN FAILURES AND THEIR COMMON REASONS

There could be many different reasons and purposes for designing engineering products. Some of those may include lowering the cost, reducing hazard, meeting competition, producing a useful item, reducing inconvenience, developing a new way, producing an item with economic worth, and meeting social changes [3–5].

In the past, even though the engineering products may have been designed with utmost care, from time to time such products have failed catastrophically or noncatastrophically. Besides well-known disasters such as the Chernobyl nuclear power reactors and the space shuttle Challenger, the other product failures include [6, 7]

- In 1937, a high school in Texas that converted from burning methane city gas to less expensive natural gas exploded, killing 455 people. A subsequent investigation revealed that the area through which the pipes ran was unsatisfactorily ventilated and recommended the installation of malodorants to all natural gas to detect leaks easily.
- In 1979, a DC-10 airplane during a flight lost an engine and crashed. An investigation of the disaster revealed that a normal engine service operation was causing the problem (i.e., as the engines were periodically dismounted, serviced, and remounted, the mounting holes became elongated during the servicing process and, thus, were the basic cause of the failure).
- In 1988, an arm on a carnival ride in Florida broke off and resulted in the death of one person and seven injuries. A subsequent investigation indicated that the arm failed because of a crack.
- In 1981, Grumman Flexible buses used in New York City developed fatigue cracks in their frames because of conflicting design restrictions and the design used a low safety factor. The design restrictions included smooth ride, low cost, fuel efficiency, light weight, accessibility for the handicapped, and maneuverability.
- In 1988, a Boeing 737-200 lost its cabin roof during flight, and a preliminary investigation discovered that several metal fatigue cracks were emanating from rivet holes in the aluminum skin.
- In 1980, in the North Sea an offshore oil rig (the Alexander L. Kielland) broke up under normal weather conditions, and a subsequent investigation revealed a 3-inch crack in a part close to a weld joint.
- In 1963, a nuclear submarine (the U.S.S. Thresher) slipped beneath the Atlantic surface and exceeded its maximum test depth and then imploded.

There could be many reasons for product failures, ranging from a disaster to a simple malfunction. Some of the common reasons include faulty reasoning, incorrect storage, incorrect usage by the consumer, erroneous data, inadequate data collection, wrong or overextended assumptions, poor understanding of the problem to be solved, wrong assembly, wrong manufacturing, wrongly stated problem with respect to basic principles, and errors in packaging and shipping.

1.3 THE DESIGN PROCESS

The design process may simply be described as an imaginative integration of scientific-related information, engineering technology, and marketing for developing a profitable product. The steps associated with the design process may vary from as few as five up to twenty-five. For example Dieter [8] described it in six steps, Vidosic [9] in eight steps, and Hill [10] in twelve steps/stages. Nevertheless, here we will describe it in six steps: (1) need recognition, (2) problem definition, (3) information gathering, (4) conceptualization, (5) evaluation, and (6) communication of design, each of which is described below.

1.3.1 NEED RECOGNITION

Prior to finding the solution to the problem, the designer or others involved with design must clearly understand and identify the needs of the user. This involves identification of customer needs, highlighting of customer needs to be satisfied, and identification of the mechanism to analyze customer needs.

1.3.2 PROBLEM DEFINITION

This requires the designer or others to develop a concise problem statement, identify requirements and limitations associated with the problem, and obtain information. This requires background investigation of the problem in question, data collection, and analysis.

1.3.3 INFORMATION GATHERING

This involves collecting information from sources such as handbooks, journals, specifications and codes, vendor catalogs, patent gazette, super information highway, technical experts, and so on.

1.3.4 CONCEPTUALIZATION

It must be remembered that creativity is a critical ingredient in finding a solution to most engineering problems. Although catalogs, textbooks, vendors, patent files, and so on are good sources for getting useful engineering ideas, there are also many idea generation techniques the engineer may also use. These include group brainstorming, synectics, morphological chart, attribute listing, and idea trigger sessions [11, 12].

1.3.5 EVALUATION

For a design engineer to decide on the best solution, all potential solutions must be compared. In addition, to perform evaluation, there must be both comparison and decision making. Typical evaluations may include the following:

- evaluation based on Go/No-Go screening
- evaluation based on feasibility judgement
- evaluation based on technology-readiness assessment

1.3.6 COMMUNICATION OF DESIGN

The final solution to the engineering problem leads to final documents representing the product or the product itself. Typically, the design documents include engineering drawings, information concerning quality assurance, bills of materials, instructions concerning operation, maintenance and retirement, and patent applications.

1.4 GOOD AND BAD DESIGN CHARACTERISTICS AND ARGUMENTS FOR AND AGAINST FINITE-LIFE DESIGN

There are various characteristics associated with good and bad designs. The characteristics of a good design include long useful life, reliability, low cost, high accuracy, simplicity, low maintenance, and attractive appearance. In contrast, the characteristics of a bad design are poor accuracy, being nonadjustable, poor reliability, rattles, rusts, cracks, being nonrepairable, high maintenance cost, short useful life, and so on.

In the design of engineering products, often various arguments are put forward for and against designing a finite-life product. Some of the arguments for a finite-life design are that high product turnover creates more jobs, reduces cost, etc.; it is useful for innovation and design development; a long-life product is not affordable; and long-life products may have to be scrapped because of continuing advancements in technology. On the other hand, arguments against the finite-life design include that it discourages good design work; it degrades the image of the product; it burdens consumers with cost of maintenance; and it discourages research and development for innovative products.

1.5 ENGINEERING DESIGN FUNCTIONS

There are many distinct functions involved in engineering design. Figure 1.1 shows five broad classifications: research-related functions, engineering-related functions, manufacturing-related functions, quality assurance–related functions, and commercial-related functions.

The research-related functions include performing basic and applied research, preparing specifications for quality testing procedures, and developing process specifications for the testing of highly stressed parts.

The engineering-related functions are the subelements of the design activity and include developing new design activity and new design concepts, production design, analyzing field problems, estimating cost, and making provisions of maintenance instructions.

The manufacturing-related functions include functions such as assembly, determining tooling requirements, cost control, purchasing materials, and manufacturing planning.

The purpose of quality assurance–related fucntions is to assure the quality of the end product, and these functions are concerned with designing methods and procedures, setting up design auditing, and so on, with respect to quality.

The commercial-related fucntions are basically concerned with relationships with various clients. Thus, the functions include performing market surveys and

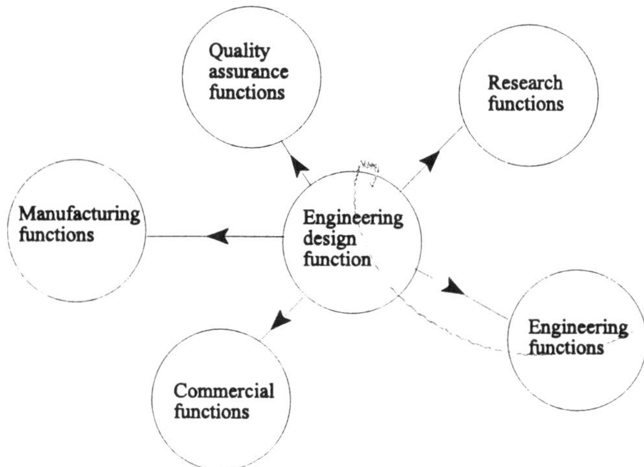

Figure 1.1 Broad classifications of engineering design functions.

tendering, managing contracts, advertising products, and arranging delivery and payments.

1.6 DESIGN REVIEWS

During the design phase of a product, various types of reviews are performed, and their sole purpose is to assure the applications of correct design principles, as well as to determine whether the design effort is progressing according to plans and specifications. Usually, the design reviews are conducted by a team of experts who examine the design under consideration from all sides. The type and size of the design review team may vary from one project to another. Furthermore, according to various experts [10], the cost of performing design reviews varies between 1% and 2% of the total engineering cost of a project [5].

Design reviews are categorized differently by different writers and practitioners, but for our purpose, we have classified them into three categories: preliminary design review, intermediate design review, and critical design review.

The preliminary design review is performed prior to the formulation of the initial design. The basic purpose of this review is to carefully examine each design specification requirement with respect to validity, completeness, and accuracy. Nevertheless, there are a number of items that should be reviewed: present/future availability of materials, potential users/customers, critical parts/components, applicable legislations, required functions, cost objective, customer requirements, schedule-imposed requirements, design alternatives, "make" or "buy" (parts), relevant constraints, and data of earlier similar products/parts [14].

The intermediate design review is performed prior to starting the detailed production drawings, and its main purpose is to compare each and every requirement of the specification with the design being developed. At this stage any change in design can still be carried out effectively. Finally, it is to be noted that prior to starting the intermediate design review, the design selection process is accomplished, and preliminary layout drawings are complete.

The critical design review is sometimes called the final design review and is performed after the completion of production drawings. At this stage a considerable amount of information is available to the design review team: test data, cost data, reports of earlier design reviews, and so on. The emphasis during this design review is on design producibility, review of analysis results, value engineering, and so on.

1.6.1 DESIGN REVIEW ITEMS AND SUBJECTS

In order to perform effective design reviews, the design review team members must have access to various items. For example, specifications and schematic diagrams, parts list, drawings, description of circuits, list of inputs/outputs, acceleration and shock test data, vibration and thermal tests data, result of failure modes and effects analysis (FMEA), and reliability predictions.

Usually during design reviews many different areas are discussed. Examples of those areas are specifications (i.e., adherence to specifications, correctness of specifications, etc.), mechanical (i.e., results of tests, thermal analysis, balance, etc.), electrical (i.e, design simplification, electrical interference, results of circuit analysis, performance, etc.), human factors (i.e., control and display, glare, labeling and marking, etc.), reproducibility (i.e., economical assembly of product in the production shop, reliance of products on a single part supplier, etc.), reliability (reliability predictions, failure modes and effect analysis, reliability allocation, etc.), maintainability (i.e., interchangeability, maintenance philosophy, etc.), safety, standardization, and value engineering [5].

1.7 ENGINEERING DESIGN MANPOWER

Usually, a team of professionals are involved in designing an engineering product. A typical design team is made up of professionals such as design engineer, reliability/quality control engineer, manufacturing engineer, tooling engineer, human factors engineer, test engineer, materials engineer, procurement engineer, and customer representatives (if applicable). In addition, the size and type of the team may vary from one project to another. The objective of each team member is to examine design from his/her speciality.

One key person of the design team is the design engineer. This individual performs many tasks, including designing the product, optimizing the design, keeping up with changing environments and technology, participating in design reviews, keeping the design within specified constraints, answering questions regarding the design, and participating in design reviews. Thus, the design engineer must possess qualities such as listed in Figure 1.2.

1.7.1 DESIGN REVIEW TEAM AND BOARD CHAIRMAN

The number and type of professionals involved in a particular design review may vary from one project to another. However, the members of a typical design review team could be designer(s), senior design engineer(s), procurement engineer, tooling engineer, manufacturing engineer, field engineer, reliability engineer, test engineer, materials engineer, quality control engineer, customer representative(s) (if applicable), and design review board chairman. Each of these individuals performs his or her design review-related functions according to their job requirement. As far as the size of the design review team is concerned, the team should not have more than 12 members for an effective performance [10].

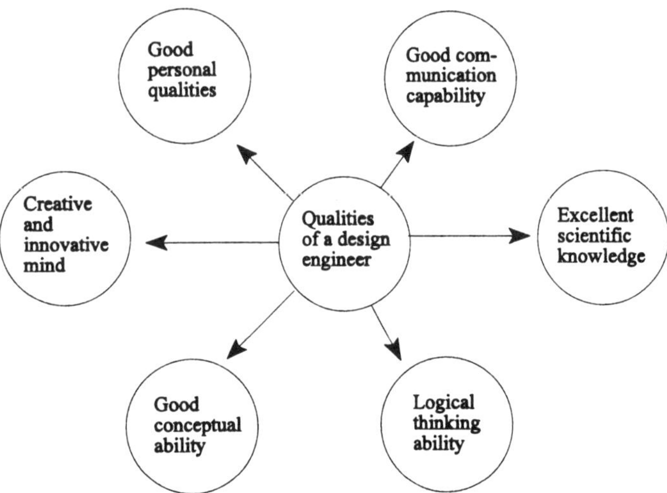

Figure 1.2 Qualities of a typical design engineer.

One key member of the design review team is the design review board chairman. This individual should belong to the engineering department but should not be in direct line of authority to the design engineer whose design is under review. His/her functions include determining and establishing the type of review to be performed, establishing the procedure for selecting particular items for review, scheduling the design reviews, chairing the meetings of the design review board, evaluating comments from the design review meetings and directing appropriate followup action(s), supervising the publication of minutes, and coordinating and providing appropriate assistance to the design organization with respect to preparation of necessary design review data.

In order to perform these functions effectively, the design review board chairman must possess certain qualities, including abroad understanding of the technical problem under consideration, appropriate skills to lead a technical meeting, freedom from bias for or against the proposed design, a high degree of tact and discretion, and a pleasant personality. In the selection of the design review board chairman, the position and the general technical competence of candidates also plays an important role. Usually, a configuration manager often heads the design review board [14].

The job of the design review board chairman is not an easy task. A good degree of carefulness is needed. In any case, some of the lead-in items to the design review are as follows [5, 10]:

- Discuss the design procedure.
- Outline design requirements.
- Provide appropriate background information.

- Identify difficulties expected or already met.
- Describe how the design meets specification requirements.

1.8 DESIGN INFORMATION SOURCES

The professionals involved with engineering design need various kinds of information that may range from past experience data on similar items to the use of the latest technology for the item currently being designed. The engineering design information sources may be grouped into two broad classifications: public or private. The public sources include federal government departments/agencies: defense, energy, mines and natural resources, transportation, science, technology and industry, etc.; local and state/provincial government departments/agencies: consumer affairs, energy, transportation, building code regulations, etc.; research establishments; universities/colleges; museums; libraries (i.e., university/college, community, etc.); and foreign missions/embassies.

Similarly, some of the private sources are nonprofit bodies/services (e.g., professional societies, trade/labor associations, and membership organizations), profit-oriented organizations (e.g., manufacturing/user organizations and consultants), individuals (e.g., university professors, colleagues/professional associates, researchers, and experts in their field of specialization). The type of information available from profit-oriented organizations include catalogs, operation and maintenance data, and cost/test data.

1.8.1 LIBRARIES

There are thousands of libraries throughout the world that could be a good source for design-related open, or unclassified, published literature. The sources of obtaining engineering design-related information in libraries may be grouped into many categories [8]: trade/professional journals, textbooks, handbooks, standards, encyclopedias, technical dictionaries, catalogs and manufacturer's brochures, technical reports, and indexing and abstracting services.

Some examples of indexing and abstracting services are Engineering Index (COMPENDEX), Sciences Citation Index (SCISEARCH), Energy Index (ENERGYLINE), International Aerospace Abstracts, Science Abstracts (INSPEC), and Applied Science and Technology Index. Selective design trade journals and books on engineering design are listed below.

Design Trade Journals

- *Product Design and Development:* This journal basically provides information on newly introduced products.

- *Machine Design:* This journal publishes information on design procedures for mechanical devices and systems, as well as a mix of conceptual and product ideas in addition to technical articles.
- *Electronic Design:* This journal is aimed for design managers and electronic design engineers and publishes information on new technology and design procedures and ideas.
- *Plastics Design Forum:* This journal is devoted to provide information on products and components in plastic.
- *Product Engineering:* This journal is primarily directed at mechanical design engineers for providing information on mechanical-related products. From time to time it also publishes general articles, thus making it attractive to electrical engineers and others.
- *Research and Development:* This journal provides information on research and developments in materials, testing and measurement, computers and software, etc.
- *Mechanical Engineering:* This journal provides practically inclined information on various areas of mechanical engineering.
- *Medical Equipment Designer:* This journal provides useful information on products, materials, and procedures concerned with the medical area.
- *Industrial Equipment News:* This journal provides information on all types of engineering equipment.
- *Military and Aerospace Electronics:* This journal provides information on advances made by the worldwide military/aerospace electronics industry.
- *New Equipment Digest:* This journal provides information on all types of engineering equipment.
- *Electronic Products:* This journal provides information on new electronic products, as well as compares products of various manufacturers.
- *Materials Engineering:* This journal basically provides information on materials for material specifiers.
- *Manufacturing Engineering:* This journal provides general information/ current development–related information in manufacturing.

Books

- Artobolevsky, I.I., *Mechanisms in Modern Engineering Design*, MIR Publishers, Moscow, 1975.
- Hill, P.H., *The Science of Engineering Design*, Holt, Rinehart & Winston, New York, 1968.
- Woodson, T.T., *Engineering Design*, McGraw-Hill Book Company, New York, 1966.
- Greenwood, D.C., *Engineering Data for Product Design*, McGraw-Hill Book Company, New York, 1982.

- Ertas, A., Jones, J.C., *The Engineering Design Process*, John Wiley & Sons, New York, 1993.
- Cross, N., *Engineering Design Methods*, John Wiley & Sons, New York, 1989.
- Dieter, G.E., *Engineering Design*, McGraw-Hill Book Company, New York, 1983.
- Dhillon, B.S., *Engineering Design: A Modern Approach*, Richard D. Irwin, Inc., Chicago, 1996.
- Greenwood, D.C., *Product Engineering Design Manual*, Krieger Publishing, Inc., Malabar, Florida, 1982.
- Love, S.F., *Planning and Creating Successful Engineering Designs*, Van Nostrand Reinhold Company, New York, 1979.
- Ullman, D.G., *The Mechanical Design Process*, McGraw-Hill Book Company, New York, 1992.
- Walton, J., *Engineering Design: From Art to Practice*, West Publishing, St. Paul, Minnesota, 1991.
- Jones, J.V., *Engineering Design: Reliability, Maintainability, and Testability*, Tab Books, Inc., Blue Ridge Summit, Pennsylvania, 1988.
- Gibson, J.E., *Introduction to Engineering Design*, Holt, Rinehart, and Winston, New York, 1968.
- Ray, M.S., *Elements of Engineering Design*, Prentice-Hall, Inc., Englewood Cliffs, New Jersey, 1985.
- Pugh, S., *Total Design: Integrated Methods for Successful Product Engineering*, Addison Wesley, Wokingham, England, 1991.
- Lewis, W., Samuel, A., *Fundamental of Engineering Design*, Prentice-Hall, Inc., Englewood Cliffs, New Jersey, 1989.
- Suh, N.P., *The Principles of Design*, Oxford University Press, New York, 1990.
- Glegg, G.L., *The Development of Design*, Cambridge University Press, New York, 1981.
- Wilcox, A.D., *Engineering Design: Project Guidelines*, Prentice-Hall, Inc., Englewood Cliffs, New Jersey, 1987.
- Hales, C., *Analysis of the Engineering Design Process in an Industrial Context*, Grants Hill Publications, East Leigh, Hants., U.K., 1991.
- French, M.J., *Conceptual Design for Engineers*, Springer-Verlag New York, Inc., New York, 1985.
- Pahl, G., Beitz, W., *Engineering Design: A Systematic Approach*, Springer-Verlag New York, Inc., New York, 1989.
- Shigley, J.E., Mischke, C.R., *Mechanical Engineering Design*, McGraw-Hill Book Company, New York, 1989.

1.9 PROBLEMS

(1) Write an essay on the history of engineering design.

(2) What are the main reasons for the failure of engineering products?

(3) Define the term *design process* and describe the typical steps associated with a design process.

(4) What are the characteristics of good and bad designs?

(5) Discuss the functions associated with engineering design.

(6) Describe the following terms:
- preliminary design review
- intermediate design review
- final design review

(7) What are the functions of a design review board chairman?

(8) Discuss the members of a typical design review team.

(9) Discuss the characteristics of a good design engineer.

(10) List arguments for and against finite-life designs.

1.10 REFERENCES

1. *Webster's Encyclopedic Dictionary*, Lexicon Publications, Inc., New York, 1988.
2. Shigley, J.D., Mitchell, L.D., *Mechanical Engineering Design*, McGraw-Hill Book Company, New York, 1983.
3. Farr, M., *Design Management*, Cambridge University Press, London, 1955.
4. Harrisberger, L., *Engineersmanship: A Philosophy of Design*, Wadsworth Publishing Company, Belmont, California, 1966.
5. Dhillon, B.S., *Quality Control, Reliability, and Engineering Design*, Marcel Dekker, Inc., New York, 1985.
6. Walton, J.W., *Engineering Design*, West Publishing Company, New York, 1991.
7. Elsayed, E.A, *Reliability Engineering*, Addison Wesley Longman, Inc., Reading, Massachusetts, 1996.
8. Dieter, G.E., *Engineering Design*, McGraw-Hill Book Company, New York, 1983.
9. Vidosic, J.P., *Elements of Design Engineering*, The Ronald Press Co., New York, 1969.
10. Hill, P.H., *The Science of Engineering Design*, Holt, Rhinehart, and Winston, New York, 1970.
11. Osborn, A.F., *Applied Imagination*, Charles Scribner and Sons, New York, 1963.
12. Dhillon, B.S., *Engineering Management*, Technomic Publishing Company, Inc., Lancaster, Pennsylvania, 1987.

13. Flurscheim, C.H., *Engineering Design Interfaces*, Design Council Publications, London, 1977.
14. AMCP 706-196, *Engineering Design Handbook, Part II: Design for Reliability*, 1976, Prepared by Headquarters, U.S. Army Material Command, 5001 Eisenhower Avenue, Alexandria, Virginia.

CHAPTER 2

Design Mathematics

2.1 INTRODUCTION

THE history of mathematics may be traced back to the origin of our present number symbols known as the Hindu-Arabic numeral system [1]. The numeral system is called Hindu-Arabic because Hindus originally invented it and the Arabs introduced it to western Europe after their invasion of the Spanish peninsula in 711 A.D. The oldest evidence of these symbols used is in India on the stone columns erected by the Scythian Indian Emperor Asoka around 250 B.C.

The first evidence of these symbols being used in the West is found in a tenth-century Spanish manuscript [1]. Nevertheless, the quadratic equations were solved by Babylonians by 2000 B.C. The beginning of the differentiation may be traced back to the ancient Greeks, but the first marked anticipation of the method of differentiation comes from ideas of Pierre Fermat (1601–1665) put forward in 1629. Laplace transforms often used to solve differential equations were the works of Pierre-Simon Laplace (1749–1827).

The earliest reference to probability is made in a gambler's guidebook written by Girolamo Cardano (1501–1576) [1]. However, the problem of dividing the winnings in a game of chance was solved correctly and independently by Blaise Pascal (1623–1662) and Pierre Fermat (1601–1665). A detailed history of mathematics is given in Reference [1] and of probability and statistics in particular in Reference [2]. This chapter presents mathematical concepts essential in understanding the contents of this book.

2.2 SET THEORY

The father of the set theory was George Ferdinand Ludwig Philip Cantor (1845–1918). A set may be described as any well-defined list, collection, or

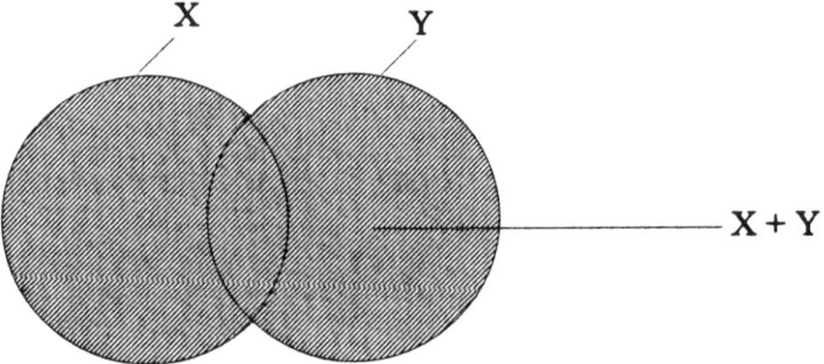

Figure 2.1 Union of sets X and Y.

class of objects. For example, objects or elements in sets could be numbers, people, rivers, and so forth. Usually, sets are denoted by capital letters, e.g., X, Y, Z, A, B, and their elements by lowercase letters, e.g., a, b, c, and d [3–5].

2.2.1 BASIC SET OPERATIONS

As in the case of operations of addition, substraction, and multiplication of numbers in regular mathematics, one is concerned with the operations of union, intersection, and difference of sets.

The *union* of sets, say X and Y, is the set of all elements that belong to X or to Y or to both. The symbol $\cup$ or $+$ is used to denote union of sets. For example, the union of X and Y is written as

$$X \cup Y \quad \text{or} \quad X + Y$$

This case is also depicted on the Venn diagram shown in Figure 2.1.

The set of all elements that belong to both sets Z and W is called the *intersection* of Z and W. The symbol $\cap$ or dot $(\cdot)$ is used to denote the intersection of sets. For example,

$$Z \cap W \quad \text{or} \quad Z \cdot W$$

The Venn diagram in Figure 2.2 shows the above case.

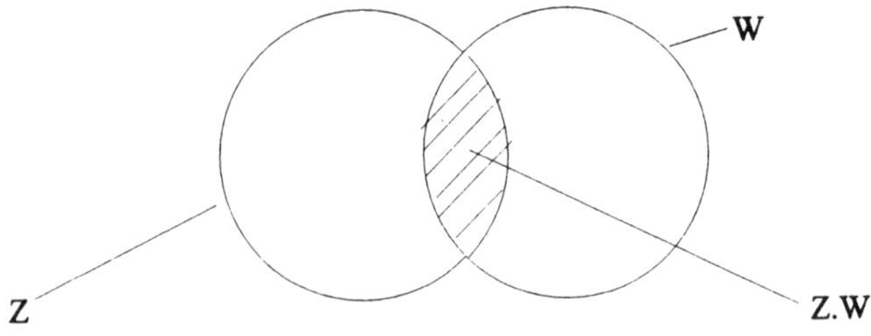

Figure 2.2 Intersection of sets Z and W.

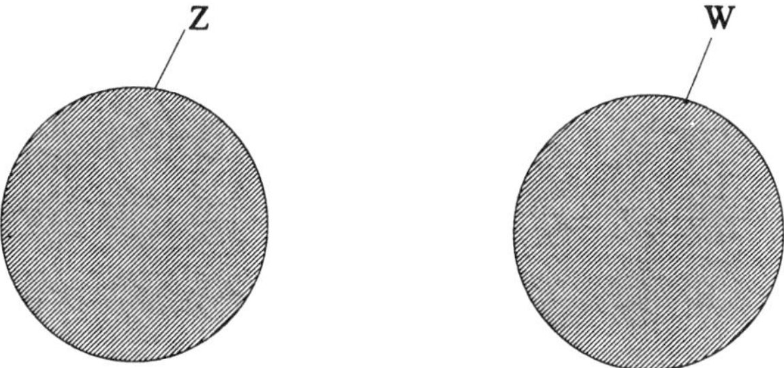

Figure 2.3 Venn diagram for mutually exclusive sets Z and W.

If the intersection of sets Z and W is zero (i.e., $Z \cdot W = 0$), the sets Z and W are called disjoint, or mutually exclusive, sets. A typical example of such a case is a fluid flow valve failing in open or closed mode. Since the valve cannot fail simultaneously in both modes, open and closed failure modes are mutually exclusive. Figure 2.3 represents mutually exclusive sets Z and W on a Venn diagram.

The sets consisting of all elements of X that do not belong to Y is called the difference of X and Y and is denoted by

$$X - Y$$

This is the shaded area in Figure 2.4.

2.2.2 LAWS OF THE ALGEBRA OF SETS

Some of the basic laws are as follows:

- idempotent laws

$$Y \cdot Y = Y \tag{2.1}$$

$$Y \quad Y = Y \tag{2.2}$$

- absorption laws

$$A + (A \cdot B) = A \tag{2.3}$$

$$A \quad (A \cdot B) = A \cdot B \tag{2.4}$$

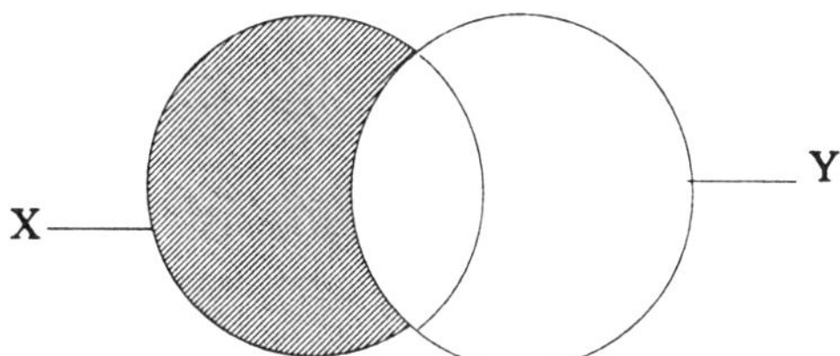

Figure 2.4 Venn diagram for the difference of sets X and Y.

- commutative laws

$$\mathrm{X} + \mathrm{Y} = \mathrm{Y} + \mathrm{X} \tag{2.5}$$

$$\mathrm{X} \cdot \mathrm{Y} = \mathrm{Y} \cdot \mathrm{X} \tag{2.6}$$

- distributive laws

$$\mathrm{A} + (\mathrm{X} \cdot \mathrm{Y}) = (\mathrm{A} \quad \mathrm{X}) \cdot (\mathrm{A} \quad \mathrm{Y}) \tag{2.7}$$

$$\mathrm{A} \cdot (\mathrm{X} \quad \mathrm{Y}) = (\mathrm{A} \cdot \mathrm{X}) + (\mathrm{A} \cdot \mathrm{Y}) \tag{2.8}$$

- associative laws

$$(\mathrm{A} + \mathrm{B}) + \mathrm{C} = \mathrm{A} + (\mathrm{B} + \mathrm{C}) \tag{2.9}$$

$$(\mathrm{A} \cdot \mathrm{B}) \cdot \mathrm{C} = \mathrm{A} \cdot (\mathrm{B} \cdot \mathrm{C}) \tag{2.10}$$

2.3 PROBABILITY

Historically, probability theory began with the study of games of chance such as cards and roulette, but today it is widely applied in engineering design. Nevertheless, the probability p of an event X may be expressed as follows: if X can occur in n ways out of N equally likely ways, then

$$p = P(X) = \frac{n}{N} \tag{2.11}$$

where $P(X)$ reads as the probability of occurrence of event X.

2.3.1 COMBINATIONS AND BINOMIAL COEFFICIENTS

In a permutation, one is concerned with the order of arrangement of the objects or elements. For example, xyz is a different permutation from yzx. However, in some cases, one may be only interested in selecting objects or elements without considering their order. Such selections are known as combinations. Thus, the number of combinations of k objects chosen from m is either denoted by ${}_mC_k$ or $\binom{m}{k}$. It is also expressed as

$$\binom{m}{k} = {}_mC_k = \frac{m!}{k!(m-k)!} \tag{2.12}$$

where

$m! = 1 \cdot 2 \cdot 3 \cdot 4 \ldots \cdot \mathrm{m}$
$k! \ = 1 \cdot 2 \cdot 3 \cdot 4 \ldots \cdot \mathrm{k}$
$o! \ = 1$

The numbers in Equation (2.12) are also called binomial coefficients because they occur in the binomial expansion. For example,

$$(a+b)^m = a^m + \binom{m}{1}a^{m-1}b + \binom{m}{2}a^{m-2}b^2 + \cdots + \binom{m}{m}b^m \qquad (2.13)$$

2.3.2 BASIC PROPERTIES OF PROBABILITY

Some of these properties are as follows [3–6]:

- The probability of occurrence of an event X is

$$0 \le P(X) \le 1 \qquad (2.14)$$

- The probability of the sample space S is

$$P(S) = 1 \qquad (2.15)$$

- The probability of negation of the sample space S is

$$P(\bar{S}) = 0 \qquad (2.16)$$

- The probability of the union of an event X and its complement $\bar{X}$ is

$$P(X + \bar{X}) = P(S) = 1 \qquad (2.17)$$

- The probability of union of k mutually exclusive events, $A_1, A_2, A_3, \ldots, A_k$, is expressed by

$$\begin{aligned} &P(A_1 + A_2 + A_3 + \cdots + A_k) \\ &\quad = P(A_1) + P(A_2) + P(A_3) + \cdots + P(A_k) \end{aligned} \qquad (2.18)$$

- The probability of union of k independent events, $A_1, A_2, A_3, \ldots, A_k$, is

$$P(A_1 + A_2 + A_3 + \cdots + A_k) = 1 - \prod_{i=1}^{k}(1 - P(A_i)) \qquad (2.19)$$

For $k = 2$, Equation (2.19) reduces to

$$\begin{aligned} P(A_1 + A_2) &= 1 - \prod_{i=1}^{2}(1 - P(A_i)) \\ &= 1 - (1 - P(A_1))(1 - P(A_2)) \\ &= 1 - [1 - P(A_1) - P(A_2) + P(A_1)P(A_2)] \\ &= P(A_1) + P(A_2) - P(A_1)P(A_2) \end{aligned} \qquad (2.20)$$

- The probability of intersection of independent events, $A_1, A_2, A_3, \ldots, A_k$, is

$$P(A_1 A_2 A_3 \cdots A_k) = \prod_{i=1}^{k} P(A_i) \tag{2.21}$$

2.3.3 CONDITIONAL PROBABILITY AND BAYES' THEOREM

The conditional probability, for example, is denoted by $P(X \mid Y)$ which reads as the probability of event X occurring, given the knowledge that event Y has occurred. In order to develop an expression for conditional probability, we assume that in k trials of an experiment, an event X occurs $[kP(X)]$ times and another event XY occurs $[kP(XY)]$ times. Thus, we see that [6]

$$kP(XY) = P(X \mid Y) kP(Y)$$

$$\therefore P(X \mid Y) = \frac{P(XY)}{P(Y)} \tag{2.22}$$

Equation (2.22) may also be stated as

$$P(X \mid Y) = \frac{\text{No. of ways } X \text{ and } Y \text{ can occur}}{\text{No. of ways } Y \text{ can occur}} \tag{2.23}$$

In the case of Bayes' theorem, we assume that $X_1, X_2, X_3, \ldots, X_m$ are mutually exclusive events whose union is the sample space S; i.e., one of the events must occur. Then if X is any event, we express Bayes' theorem as follows [7]:

$$P(X_i \mid X) = \frac{P(X_i) P(X \mid X_i)}{\sum_{i=1}^{m} P(X_i) P(X \mid X_i)} \tag{2.24}$$

Often, Bayes' theorem is referred to as a theorem on the probability of causes because it allows us to find the probabilities of various events such as X_1, X_2, $X_3, \ldots, X_m$ that can cause X to occur.

2.4 PROBABILITY DENSITY FUNCTION BASIC PROPERTIES AND DEFINITION

Two basic properties of a probability density function, $f(t)$, are as follows:

$$f(t) \geq 0 \tag{2.25}$$

where t is the random variable, and

$$\int_{-\infty}^{\infty} f(t)\,\mathrm{d}t = 1 \tag{2.26}$$

Equation (2.26) states that the total area under a probability density function is equal to unity.

A probability density function may be defined as

$$f(t) = \frac{\mathrm{d}F(t)}{\mathrm{d}t} \tag{2.27}$$

where

$$F(t) = \int_{-\infty}^{t} f(x)\,\mathrm{d}x \tag{2.28}$$

and

$$F(\infty) = 1 \tag{2.29}$$

$F(t)$ is known as the cumulative distribution function of the continuous random variable t.

2.4.1 EXAMPLE 2.1

Assume that the times to failure of an electric fan are defined by

$$f(t) = \lambda e^{-\lambda t} \tag{2.30}$$

where

t = time
λ = the electric fan failure rate
$f(t)$ = the failure or probability density function

Prove that the total area under Equation (2.30) curve is equal to 1.

Substituting Equation (2.30) into Equation (2.26) we get

$$\begin{aligned} F(\infty) &= \int_{-\infty}^{\infty} \lambda e^{-\lambda t}\,\mathrm{d}t \\ &= \int_{0}^{\infty} \lambda e^{-\lambda t}\,\mathrm{d}t \\ &= \left[\frac{\lambda e^{-\lambda t}}{-\lambda}\right]_0^{\infty} \\ &= 1 \end{aligned}$$

Thus, it is proven that the total area under Equation (2.30) curve is equal to unity.

2.5 MEAN VALUE AND VARIANCE OF A CONTINUOUS RANDOM VARIABLE

The mean or expected value, μ, of a continuous random variable is defined by

$$\mu = \int_{-\infty}^{\infty} t f(t)\,\mathrm{d}t \tag{2.31}$$

The variance, $\sigma^2(t)$, of a random variable t is given by

$$\sigma^2(t) = -2 \lim_{s \to 0} R'(s) - \mu^2 \tag{2.32}$$

where

$s =$ the Laplace transform variable
$R'(s) =$ the derivative of $R(s)$ with respect to s
$R(s) =$ the Laplace transform of $1 - F(t)$.

2.5.1 EXAMPLE 2.2

Find the mean value of Equation (2.30). Substituting Equation (2.30) into Equation (2.31), we get

$$\begin{aligned} \mu &= \int_{-\infty}^{\infty} t\lambda e^{-\lambda t}\,\mathrm{d}t \\ &= \lambda \int_{0}^{\infty} t e^{-\lambda t}\,\mathrm{d}t \\ &= \lambda \left[\frac{e^{-\lambda t}}{-\lambda} \left(t + \frac{1}{\lambda} \right) \right] \Bigg|_0^{\infty} \\ &= \frac{1}{\lambda} \end{aligned} \tag{2.33}$$

Thus, the mean value of Equation (2.30) is $1/\lambda$.

2.6 PROBABILITY DISTRIBUTIONS

Probability distributions play an important role in performing analysis of engineering systems. Probability distributions are used to represent, for example, failure behavior of various items/systems. There are a large number of different probability distributions used in analyzing engineering systems. In particular, some of the probability distributions used during system design are exponential: Rayleigh and Weibull, especially in evaluating systems from the reliability aspect.

2.6.1 EXPONENTIAL DISTRIBUTION

In reliability analysis, this distribution is often used to represent times to failure of various engineering items [8]. The probability density function of the distribution is defined by

$$f(t) = \lambda e^{-\lambda t} \qquad t \geq 0, \lambda > 0 \tag{2.34}$$

where

t = time
λ = the distribution parameter

By substituting Equation (2.34) into Equation (2.28), we get the following expression for the cumulative distribution function:

$$\begin{aligned} F(t) &= \int_0^t \lambda e^{-\lambda x} \mathrm{d}x \\ &= \left[\frac{\lambda e^{-\lambda x}}{-\lambda}\right]_0^t \\ &= 1 - e^{-\lambda t} \end{aligned} \tag{2.35}$$

Substituting Equation (2.34) into Equation (2.31) leads to the following expression for the mean value:

$$\begin{aligned} \mu &= \lambda \int_0^\infty t e^{-\lambda t} \mathrm{d}t \\ &= \frac{1}{\lambda} \end{aligned} \tag{2.36}$$

Example 2.3 Assume that the times to failure of an electric motor are exponentially distributed; thus, its failure rate is 0.0005 failures per hour. Calculate the probability of the electric motor failing during a 100-hour mission. Substituting the given data into Equation (2.35) yields

$$\begin{aligned} F(100) &= 1 - e^{-(0.0005)(100)} \\ &= 0.0487 \end{aligned}$$

The probability of the electric motor failing is 0.0487.

2.6.2 RAYLEIGH DISTRIBUTION

Rayleigh distribution is often used in the theory of sound. In reliability studies, it is used to represent failure times of items whose failure rates are linearly increasing. The probability density function of the Rayleigh distribution is

$$f(t) = \frac{2}{\theta} t e^{-t^2/\theta}, \qquad t \geq 0, \theta > 0 \tag{2.37}$$

where θ is the distribution scale parameter.

The following cumulative distribution function of the distribution is obtained by substituting Equation (2.37) into Equation (2.28):

$$\begin{aligned} F(t) &= \frac{2}{\theta} \int_0^t x e^{-x^2/\theta} \mathrm{d}x \\ &= 1 - e^{-t^2/\theta} \end{aligned} \tag{2.38}$$

The mean or expected value of the distribution is obtained by substituting Equation (2.37) into Equation (2.31):

$$\begin{aligned} \mu &= \frac{2}{\theta} \int_0^\infty t^2 e^{-t^2/\theta} \mathrm{d}t \\ &= (\sqrt{\pi\theta})/2 \end{aligned} \tag{2.39}$$

2.6.3 WEIBULL DISTRIBUTION

This distribution was developed by W. Weibull [9] in the early 1950s and is used to represent many different physical phenomena. The probability density function of the distribution is defined by

$$f(t) = \frac{1}{\theta} t^{a-1} e^{-t^a/\theta}, \qquad \theta, a > 0, t \geq 0 \tag{2.40}$$

Substituting Equation (2.40) into Equation (2.28), we get the following expression for the distribution cumulative distribution function:

$$\begin{aligned} F(t) &= \frac{1}{\theta} \int_0^t t^a e^{-t^a/\theta} \\ &= 1 - e^{-t^b/\theta} \end{aligned} \tag{2.41}$$

The following expression for mean or expected value of the distribution is obtained by substituting Equation (2.41) into Equation (2.31):

$$\begin{aligned}\mu &= \frac{1}{\theta}\int_0^{\infty} t^a e^{-t^a/\theta} dt \\ &= (\theta)^{1/a}\Gamma\left(1+\frac{1}{a}\right)\end{aligned} \tag{2.42}$$

where $\Gamma(\cdot)$ is the gamma function.

Example 2.4 Prove that for $a = 1$ and 2, the cumulative distribution function of the Weibull distribution is the same as that of the exponential and Rayleigh distributions, respectively. Thus, substituting $a = 1$ and 2 in Equation (2.41), we get

$$F(t) = 1 - e^{-t/\theta} \tag{2.43}$$

and

$$F(t) = 1 - e^{-t^2/\theta} \tag{2.44}$$

For $\theta = 1/\lambda$ Equation (2.43) is the same as for the exponential distribution, i.e., Equation (2.35). Also, Equation (2.44) is the same as for the Rayleigh distribution, i.e., Equation (2.38).

2.7 LAPLACE TRANSFORMS

In design work Laplace transforms are often used to solve a system of linear differential equations with constant coefficients, especially in reliability evaluation of newly designed systems. Laplace transform of a function $f(t)$ is defined by

$$f(s) = \int_0^{\infty} e^{-st} f(t) dt \tag{2.45}$$

where

t = time
s = the Laplace transform variable

Laplace transforms of various functions are presented in Table 2.1.

TABLE 2.1. Laplace Transforms of Selected Functions.

$f(t)$	$f(s)$
k, for $k = 1, 2, 3, \ldots,$	$\frac{k}{s}$
$e^{-\lambda t}$	$\frac{1}{s+\lambda}$
$\frac{t^{m-1}}{(m-1)!}$, for $m = 1, 2, 3, \ldots,$	$\frac{1}{s^m}$
$\frac{\mathrm{d}\, f(t)}{\mathrm{d}t}$	$sf(s) - f(0)$

2.7.1 EXAMPLE 2.5

Obtain Laplace transforms of functions t^2 and $e^{-\lambda t}$, where λ is a constant. Substituting t^2 for $f(t)$ in Equation (2.45), we get

$$\begin{aligned} f(s) &= \int_0^{\infty} t^2 e^{-st}\,\mathrm{d}t \\ &= \frac{e^{-st}}{-s}\left(t^2 + \frac{2t}{s} + \frac{2}{s^2}\right)\Bigg|_0^{\infty} \\ &= \frac{2}{s^3} \quad \text{for} \quad s > 0 \end{aligned} \tag{2.46}$$

Similarly, substituting $e^{-\lambda t}$ for $f(t)$ in Equation (2.45) yields

$$\begin{aligned} f(s) &= \int_0^{\infty} e^{-st} e^{-\lambda t}\,\mathrm{d}t \\ &= \frac{e^{-(s+\lambda)t}}{-(s+\lambda)}\Bigg|_0^{\infty} \\ &= \frac{1}{s+\lambda} \end{aligned} \tag{2.47}$$

2.8 FINAL VALUE THEOREM

This is expressed as follows:

$$\lim_{t \to \infty} f(t) = \lim_{s \to 0} sf(s) \tag{2.48}$$

Equation (2.48) may be proved by using the relation for the Laplace transform of a derivative given in Table 2.1 [10, 11]:

$$\int_0^{\infty} e^{-st} \frac{\mathrm{d}f(t)}{\mathrm{d}t} \cdot \mathrm{d}t = sf(s) - f(0) \tag{2.49}$$

The limit of the left-hand side of Equation (2.49) as s approaches zero is

$$\begin{aligned}\lim_{s\to 0}\int_0^\infty e^{-st}\frac{\mathrm{d}f(t)}{\mathrm{d}t}\mathrm{d}t &= \int_0^\infty \frac{\mathrm{d}f(t)}{\mathrm{d}t}\mathrm{d}t = \lim_{z\to\infty}\int_0^z \frac{\mathrm{d}f(t)}{\mathrm{d}t}\mathrm{d}t \\ &= \lim_{z\to\infty}[f(z)-f(0)] \\ &= \lim_{t\to\infty} f(t)-f(0)\end{aligned} \tag{2.50}$$

Similarly, the limit of the right-hand side of Equation (2.49) as s approaches zero is

$$\lim_{s\to 0} sf(s) - f(0) \tag{2.51}$$

Thus, equating Relationships (2.50) and (2.51), we get

$$\begin{aligned}\lim_{t\to\infty} f(t) - f(0) &= \lim_{s\to 0} sf(s) - f(0) \\ \lim_{t\to\infty} f(t) &= \lim_{s\to 0} sf(s)\end{aligned} \tag{2.52}$$

This proves that Equation (2.52) is identical to Equation (2.48).

2.8.1 EXAMPLE 2.6

Using the left- and right-hand sides of Equation (2.48), prove that the steady-state value of the following function is the same:

$$f(t) = \frac{\lambda_1}{\lambda_2+\lambda_1} + \frac{\lambda_2}{\lambda_2+\lambda_1}e^{-(\lambda_1+\lambda_2)t} \tag{2.53}$$

where λ_1 and λ_2 are the constants.

Taking the Laplace transform of Equation (2.53), we get

$$f(s) = \frac{(s+\lambda_1)}{s(s+\lambda_1+\lambda_2)} \tag{2.54}$$

Substituting Equation (2.54) into the right-hand side of Equation (2.52) leads to

$$\lim_{s\to 0}\frac{s(s+\lambda_1)}{s(s+\lambda_1+\lambda_2)} = \frac{\lambda_1}{\lambda_1+\lambda_2} \tag{2.55}$$

Similarly, substituting Equation (2.53) into the left-hand side of Equation (2.52), we get

$$\lim_{t\to\infty}\left[\frac{\lambda_1}{\lambda_2+\lambda_1}+\frac{\lambda_2}{\lambda_2+\lambda_1}e^{-(\lambda_1+\lambda_2)t}\right]=\frac{\lambda_1}{\lambda_2+\lambda_1} \tag{2.56}$$

The right-hand sides of Equations (2.55) and (2.56) are the same. Thus, it proves that the left- and right-hand sides of Equation (2.48) lead to the same end result.

2.9 SOLVING DIFFERENTIAL EQUATIONS USING LAPLACE TRANSFORMS

One of the most powerful applications of Laplace transforms is to find the solution to linear differential equations with constant coefficients in reliability analysis. The following example demonstrates the application of Laplace transforms to solve a differential equation.

2.9.1 EXAMPLE 2.7

The following differential equation describes a system composed of n independent units in series (if any one of the units fails, the series system fails):

$$\frac{dP(t)}{dt}+\sum_{i=1}^{n}\lambda_i P(t)=0$$
$$\text{At time } t=0,\ P(0)=1 \tag{2.57}$$

where

λ_i = the constant failure rate of unit i; for $i=1,2,3,\ldots,n$
$P(t)$ = the probability of success of the series system at time t

Solve the differential equation by using Laplace transforms. Taking Laplace transforms of Equation (2.57) leads to

$$sP(s)-P(0)+\left(\sum_{i=1}^{n}\lambda_i\right)P(s)=0 \tag{2.58}$$

Since $P(0)=1$, Equation (2.58) is rearranged to the following form:

$$P(s)=\frac{1}{s+\sum_{i=1}^{n}\lambda_i} \tag{2.59}$$

Taking the inverse Laplace transform of Equation (2.59) results in

$$P(t) = e^{-\sum_{i=1}^{n} \lambda_i t} \tag{2.60}$$

Equation (2.60) is the solution to Equation (2.57).

2.10 PROBLEMS

(1) Write an essay on the history of the probability theory.

(2) Prove the distributive law of the boolean algebra.

(3) Prove that the probability of union of n independent events, $x_1, x_2, \ldots, x_n$ is

$$P(x_1 + x_2 + \cdots + x_n) = 1 - \prod_{i=1}^{n}(1 - P(x_i)) \tag{2.61}$$

(4) Prove that the total area under the Weibull probability density function curve is equal to unity.

(5) Obtain an expression for variance associated with the Rayleigh distribution.

(6) Take Laplace transforms of the following function:

$$f(t) \frac{\lambda_2}{\lambda_2 + \lambda_1} - \frac{\lambda_2}{\lambda_2 + \lambda_1} e^{-(\lambda_1 + \lambda_2)t} \tag{2.62}$$

where λ_1 and λ_2 are the constants.

(7) Prove that the variance, $\sigma^2(t)$, of a random variable t is expressed by

$$\sigma^2(t) = -2 \lim_{s \to 0} R'(s) - \mu^2 \tag{2.63}$$

where

s = the Laplace transform variable
$R(s)$ = the Laplace transform of $[1 - F(t)]$
$R'(s)$ = the derivative of $R(s)$ with respect to s.

(8) Solve the following system of differential equations using Laplace transforms:

$$\frac{\mathrm{d}P_0(t)}{\mathrm{d}t} + (\lambda_1 + \lambda_2)P_0(t) = 0 \tag{2.64}$$

$$\frac{\mathrm{d}P_1(t)}{\mathrm{d}t} - \lambda_1 P_0(t) = 0 \tag{2.65}$$

$$\frac{\mathrm{d}P_2(t)}{\mathrm{d}t} - \lambda_2 P_0(t) = 0 \tag{2.66}$$

At time $t = 0$, $P_0(0) = 1$, $P_1(0) = P_2(0) = 0$, where

$P_i(t)$ = the probability of system being in state i at time t; for $i = 0, 1, 2$.
λ_i = the ith constant; for $i = 1, 2$.

(9) Find the Laplace transforms of the following functions:
- t^k; for $k = 1, 2, 3, \ldots$
- $\frac{d^2 f(t)}{dt^2}$

2.11 REFERENCES

1. Eves, H., *An Introduction to the History of Mathematics*, Holt, Rinehart and Winston, New York, 1976.
2. Owen, D.B., Editor, *On the History of Statistics and Probability*, Marcel Dekker, Inc., New York, 1976.
3. Lipschutz, S., *Theory and Problems of Probability*, McGraw-Hill Book Company, New York, 1965.
4. Spiegel, M.R., *Theory and Problems of Probability and Statistics*, McGraw Hill Book Company, New York, 1975.
5. Lipschutz, S., *Set Theory and Related Topics*, McGraw-Hill Book Company, New York, 1964.
6. Spiegel, M.R., *Probability and Statistics*, McGraw-Hill Book Company, New York, 1975.
7. Ramakumar, R., *Engineering Reliability*, Prentice-Hall, Englewood Cliffs, New Jersey, 1993.
8. Davis, D.J., An Analysis of Some Failure Data, *J. Amer. Stat. Assoc.*, June 1952, pp. 113–150.
9. Weibull, W., A Statistical Distribution of Wide Applicability, *J. Appl. Mech.*, Vol. 18, 1951, pp. 293–297.
10. Spiegel, M.R., *Theory and Problems of Laplace Transforms*, McGraw-Hill Book Company, New York, 1965.
11. Nixon, F.E., *Handbook of Laplace Transformation*, Prentice-Hall, Inc., Englewood Cliffs, New Jersey, 1965.

CHAPTER 3

Concurrent Engineering, Reverse Engineering, and Reengineering

3.1 INTRODUCTION

TODAY, design engineers are expected to be expert in many disciplines that may range from manufacturing to human factors. Since this does not look very realistic, the team approach to designing is an alternative option. Because of the decrease in the United States' industry's market share in the 1980s, the Ford Motor Company practiced the team, or concurrent, engineering approach during the design and development of its Taurus model [1]. In 1982, a study was initiated by the U.S. Defense Advanced Research Projects Agency (DARPA) to develop ways and means to enhance concurrency in the product design process [2], and it was not until 1986 that the term *Concurrent Engineering* was used for the first time in a report (R-338) prepared by the Institute for Defense Analyses (IDA). Another term used for concurrent engineering is "simultaneous engineering." There are various definitions used to describe concurrent engineering, one of which is the following: "Concurrent engineering is the simultaneous, interactive, and interdisciplinary involvement of professionals belonging to areas such as design, manufacturing and field support to decrease product development cycle time while ensuring factors such as performance, reliability, quantity, and support responsiveness" [3].

Japanese companies are appeared to be leaders in implementing concurrent engineering design and have succeeded in reducing by half the time to deliver major products such as aircraft and automobiles in comparison to that of the U.S. companies [3]. However, in the past many U.S. companies have employed concurrent engineering and reported various successes; e.g., John Deere & Co. reduced the cost of developing new construction equipment by 30% and 60% in development time, AT&T Co. succeeded in reducing 50% of the time to develop a 5ESS electronic switching system, and the Boeing company predicted releasing design drawings of its giant 777 transport aircraft a year and a half earlier in comparison to that of its 767 plane [4].

The concept of reverse engineering could be practiced to eradicate weaknesses in or to enhance the capabilities of existing facilities. Reverse engineering may be expressed as the act of creating specifications for a facility by individuals other than the designers of that facility, basically based upon analysis and dimensioning a specimen or specimens [5]. It appears that the modern reverse engineering concept had its beginning in the 1980s; for example, a U.S. Government Acquisition Regulation in the mid-1980s categorically stated that the application of the reverse engineering concept should only be considered when economic feasibility justifies it and all other options for developing such technical data are not an attractive proposition [6]. Furthermore, to maintain a competitive advantage on General Motors Corporation products, the Ford Motor company (and vice versa) practiced it in the 1980s [5].

In the 1990s, the term *reengineering* is often used in industrial and other sectors. Reengineering may simply be described as the examination and alteration of a system or an item under consideration to reconstitute it in a new form and for the ultimate implementation of that form [7]. Usually, reengineering includes some form of reverse engineering (i.e., to attain a more abstract description) followed by restructuring or some kind of forward engineering. This chapter describes concurrent engineering, reverse engineering, and reengineering.

3.2 CONCURRENT ENGINEERING

The importance of the application of concurrent engineering arises from factors such as the design activity having a significant impact on the item production testing and servicing costs. In other words, the decisions made during the design phase of an item will have the biggest impact on the overall cost of that item during its life span. In order to compete effectively in today's global market, the cost associated with all elements of product design must be reduced. Factors such as cost have led many companies [8] to practice concurrent engineering: Hewlett-Packard, Sun Microsystems, Mercury Computer Systems, and Digital Equipment Corporation.

The cornerstone of concurrent engineering is a team effort in which professionals belonging to different disciplines work hand in hand during the development of new products. Obviously, this new concept of executing tasks in parallel is generally faster compared to performing those tasks sequentially. However, for concurrent engineering to succeed in an organization, a significant cultural change is necessary because professionals from many disciplines will be involved, for example, manufacturing, maintenance, marketing, finance, accounting, and customers.

3.2.1 CONCURRENT ENGINEERING OBJECTIVES

In fact, strictly speaking, the objective of practicing concurrent engineering in an organization is not solely cost reduction; there could be many others. For example, in the 1980s Hewlett-Packard Co. decided to improve the quality of its manufactured products by 100% in 5 years [9] and succeeded without any difficulty [10]. Since the application of concurrent engineering could have various objectives, some of the principal ones are as follows [11]:

- Reduce product development costs. These costs are generally an important component in the selling price of a product. Reduction in such costs may improve the competitive advantage of the manufacturer.
- Reduce testing costs. These costs are becoming a larger component of the overall product cost equation as the many manufactured products continue to increase in complexity and sophistication. One example of increase in the testing costs in such products is the incorporation of built-in test systems
- Lower service costs. These costs are often large for big products in many organizations because such products usually require some degree of servicing after their installation at the customers' facility. Concurrentengineering helps to direct attention during product design to reduce serviceability associated costs.
- Reduce manufacturing costs. These costs are a significant component of the total product cost. Concurrent engineering is a useful tool to reduce manufacturing costs by producing manufacturing friendly product designs.
- Reduce marketing time. In order to be competitive in today's market, the shorter time to market is probably the most widely used phrase. It basically means responding faster to the customer requirements, and concurrent engineering is a useful concept in this aspect.
- Improve product quality. Today's customer is very quality conscious; thus, the better quality products are likely to take a greater share of the market. In many ways the application of the concurrent engineering helps to improve product quality. In fact, the main objective of practicing concurrent engineering in some organizations is to improve their product quality. For example, Hewlett-Packard Co. used the concurrent engineering concept to improve the quality of its products by 100% in a specified time period [9].
- Improve competitiveness of manufactured products. As the competition increases on a global scale, many manufactured products have to compete against internally and externally manufactured products. Concurrent

engineering is not only practiced for increasing the market share, but also to maintain market share because the competitors could very well be taking advantage of it.

- Increase profit margins. Because profit margin is very important to organizations in business, many companies employ concurrent engineering to improve their profit margins.

3.2.2 APPLYING THE CONCURRENT ENGINEERING CONCEPT

The application of the concurrent engineering approach may not be suitable to all organizations. Factors such as product type, organization size, process followed to produce a product, and physical locations of the company branches may dictate the degree and the need of concurrent engineering. In any case, to develop an overall vision of a concurrent engineering environment in an organization, attention must be given to the following areas [2]:

- product development environment
- current standing
- envisioned objective achievement methods
- plans to achieve set objectives

The product development environment is concerned with the assessment action. In other words, it deals with assessing an organization's existing product development environment. Available tools for this purpose are the Software Engineering Institute Assessment Questionnaire, the Department of Defense CALS/Concurrent Engineering Task Group for Electronic Systems Self Assessment, and the Mentor Graphics Corporation Process Maturity Assessment Questionnaire [2]. These tools will be helpful to determine the concerned company's current product development environment with respect to concurrent engineering; the company's appropriate concurrent engineering vision; necessary, required actions to implement the vision for balancing each existing and expected concurrent engineering dimension; and priorities for implementing the company's concurrent engineering vision.

The information on the current standing of the organization to develop its products is quite useful to plan for future actions. In other words, this stage is basically concerned with determining the organization's current standing with respect to product development. The company assessment questionnaire is a useful tool for this purpose. Its four main areas for collecting information are as follows:

- organization: The questionnaire specifically addresses questions on team integration, empowerment, automation support, and training and organization.

- communication infrastructure: Because the concurrent engineering emphasizes the simultaneous functioning of various groups, the effective functioning of the communication infrastructure is quite crucial. The three main areas for collecting information are product development, product data, and feedback.
- requirements: The effectiveness in meeting internal and external requirements is vital for the successful application of concurrent engineering. The main categories for collecting requirement-related information are the requirement's definition, planning methodology, planning perspective, validation, and standards.
- product development: The assessment questionnaire collects information on areas such as component engineering, design process, and optimization.

In the case of envisioned objective achievement methods, the objective is to determine the approaches that can be used in product development environments to attain the full benefits of concurrent engineering. The methods matrix approach is a useful tool for this purpose and is described in detail in Reference [2].

For achieving a successful concurrent engineering environment, careful planning to attain set objectives is required. The analysis of the company assessment questionnaire and the methods matrix provides a useful input to develop attainment plans. More specifically, actions such as listed below are necessary:

- assessment result plotting and analysis
- communicating the message with other involved individuals
- refining the concurrent engineering vision
- validating achievements with plans
- making improvements continuously

During the consideration for the application of the concurrent engineering concept, attention must be paid to the following four key factors [12]:

- concurrence
- constraints
- coordination
- consensus

The concurrence factor simply reemphasizes the main objective of concurrent engineering; i.e., product and process design is carried out during the same time period. The constraints factor is concerned with considering the capabilities and limitations of available manufacturing processes during the design phase and ensuring that the product design is compatible with them. The coordination factor deals with closely coordinating the product and process requirements and other associated objectives during the design phase. The consensus factor calls for the participation of the full concurrent engineering team and its unanimity on important product design-related decisions.

Concurrent Engineering Approach Introduction-Related Factors
It is not that simple to introduce the concurrent engineering approach in an organization. A careful consideration and groundwork is necessary. It is advisable to seek answers to general factors such as listed below when contemplating concurrent engineering introduction [12]:

- starting date of the concurrent engineering activity
- location of the concurrent engineering activity
- approach to be followed to manage the concurrent engineering team
- degree of reliance on external (i.e., outside the company) expertise with respect to concurrent engineering
- procedure to be followed in evaluating concurrent design project results
- degree of training required for the team members (to able) to work as a group
- reporting of the team within the company
- team members' physical location
- ways and means to be followed in weighing conflicting objectives
- ways and means to be followed to ensure team productivity
- management's expectations with respect to project

3.2.3 CONCURRENT ENGINEERING TEAM

The use of teams is the cornerstone of the concurrent engineering concept, and their formation is not an easy task and requires careful consideration. Before going on to the subject of concurrent engineering teams, let us define the words *team* and *teamwork*. According to *Webster's Encyclopedic Dictionary* [13], *team* and *teamwork* are defined, respectively, as "a number of people work together on common task" and "the quality whereby individuals unselfishly subordinate their own part to the general effort of the group with whom they are working or playing." In the industrial sector these definitions partially cover the real meanings of the team and teamwork because there could be contributors other than the team members in addition to the dependence of the team decisions on factors such as budget, resources, and so on. The athletic teams are probably close to the simple understanding of the teams and teamwork. In order to understand the difference between the athletic and industrial teams, we compare important factors associated with them: clear rules, fair play, ability to self-manage, team goal/measurement/timeliness, roles, methods, competition, and leader [14]. The corresponding applicable factors associated with the athletic team (in parentheses for the industrial team) are rule book (undefined), penalties/referees (seldom called/never watched), self-contained (dependent on others), win/points/immediately known (win/sales/long-term), defined/practice/extensive training (assigned/unclear/little training), practice/signals for coordination (selected/ad hoc coordination), scouted/filmed/planned attack (opinion/trade information), and coach (team member).

Team Formation

Team formation is an important element in the success of concurrent engineering concept application. It is not only important to include members from important disciplines or areas, but they must have the proper qualifications, experiences, interpersonal skills, and so forth. A typical team includes members such as those listed below [12, 14].

- concurrent engineering mentor
- team leader (engineering)
- marketing manager
- engineering manager (lead)
- design engineer
- manufacturing engineer
- information technology specialist
- service engineer
- software engineer
- quality control engineer
- reliability engineer
- safety engineer
- vendor/customer representative
- human factor/environmental specialist

According to a 1990 study, the typical component design team (electronic industry) in the United States was made up, on the average, of eight individuals, as opposed to eighteen in Japan [15]. Some of the important team characteristics are that members represent all concerned disciplines, possess the capability for compromising and accepting consensus decisions, possess a broad view, are dedicated to the success of the company, possess some knowledge about other disciplines, and have a strong team spirit.

Team Charter

For the success of the concurrent engineering application, it is useful to have a team charter. The charter covers the member's function and authority, as well as helps to avoid misunderstanding or confusion. In any case, some of the items that the charter normally includes are product description, team objective, competitive situation description, team goals, team member's role, support from the company president and/or other senior management people, team membership, team authority boundary, accountability of the team to a senior manager, budget authority, implementation requirements, decision making, and level of decision making.

Team Plan

Usually, the team plan includes the important components of the activity. The major categories that should be included at the initial stage of the product development cycle are team membership and charter, important competing products, customer need projections and market requirements, applicable standards, projected product volume and cost, manufacturing risks, preliminary development schedule, customer information collection process, important competitive advantages, key technologies to be used, competitor's marketing strategy projection, product distribution needs, development budget requirements, regulatory requirements, and preliminary business projects.

Team Leadership and Management

Just like in the case of any other engineering team, the function of the concurrent engineering leader is very important. Therefore, a careful consideration must be given in appointing that leader. The leader can either come from the team members or the next level managers. The leader should possess the following traits and responsibilities [15]:

- ability to resolve issues through consensus
- capability of recognizing that all individuals are different
- memory and wisdom to avoid repeating the same mistakes
- ability to motivate team members
- tactfulness to lead people
- flexibility
- ability to reason
- good interpersonal skills
- knowledge to garner necessary resources such as human, hardware, and software
- strong participation in defining the problems to be addressed
- ability to empower team members to make decisions and settle disputes
- ability to assign tasks to team members

- aptitude to set the agenda for each meeting
- ability to ensure the distribution of necessary information to all concerned in a timely manner

Obviously, the main purpose of having a good concurrent engineering team leader is to manage the team members effectively. Some suggestions to manage the team in an effective manner are as follows [14]:

- The team should perform its function under its own leadership.
- Maximize team collaboration.
- Establish goals for the team to achieve.
- Maximize organizational support.
- Review and measure the team contribution.
- Hold team meetings regularly.
- Brainstorm as appropriate among team members to overcome difficulties.
- Prepare team meeting agendas as clear as possible and distribute among all concerned people 3 to 5 days prior to the meeting.
- Record minutes of meetings and distribute to all concerned people soon after each meeting.
- Develop trust among team members.

3.2.4 METHODOLOGIES AND TECHNIQUES RELATED TO THE CONCURRENT ENGINEERING PROCESS

There are many methodologies and techniques that are useful in the concurrent engineering process. This section briefly describes some of these approaches [14, 16].

- quality function deployment (QFD): This is also known as the "house of quality" and was first applied by the two Japanese giants: Mitsubishi Corporation and Toyota Motor Corporation. QFD may simply be described as a pair of spreadsheets relating subjective customer's desires known as customer attributes to quantitative engineering characteristics (EC). In simple words, QFD is a method of rating the importance of certain product features by considering customer inputs and then comparing such features to competitive products. This approach provides a justification for establishing important product features in addition to the projected product price. QFD is used during the conceptual phase of the design process.
- design for manufacturing: This is basically concerned with understanding the product's future manufacturing processes during the design phase and

then taking appropriate measures. This helps to reduce manufacturing costs and to improve manufacturing quality.

- robust design: This concept was developed by Genichi Taguchi during 1949–1961 at the Electrical Communications Laboratory of Nippon Telegraph and Telephone Company, Japan. The approach aims to minimize less-than-optimal interactions among a product's elements, which are caused by external factors such as the environment, abusive operation, and manufacturing process variations.
- fishbone diagram: The reason for calling this approach "fishbone diagram" is because of its resemblance to a fish's spine-and-rib structure, which is particularly applicable to concurrent engineering. This approach begins with an effect as a spine and works backward, with major causes added as ribs.
- experimental design approaches: There are various available experimental design analysis techniques developed by Genichi Taguchi and others that are useful for examining the effect of changing a single factor while holding the other factors constant.
- Pugh process [17]: This is a useful analysis method of comparing alternatives, in particular determining the better attributes out of the various alternative attributes. Furthermore, the process is a useful tool for the design team to gain a deeper look at the strengths and weaknesses of all possible potential solutions.

3.2.5 GENERAL CONCURRENT ENGINEERING GUIDELINES, CHARACTERISTICS OF THE EFFECTIVE CONCURRENT ENGINEERING TEAM, AND HANDLING CONFLICTS IN THE CONCURRENT ENGINEERING ENVIRONMENT

Over the years various individuals associated with concurrent engineering have developed sets of useful guidelines. For example, Turino [11] has called the following the concurrent engineering commandments:

- Establish a multidisciplinary design team.
- Communicate effectively with the product customer/user.
- Involve outside people such as suppliers and subcontractors at the early stage of the project.
- Conduct simulation analysis of product/process performance(s).
- Design associated processes concurrently with the product.
- Integrate technical reviews effectively.
- Take advantage of "lessons learned/experience gained."

- Integrate computer-aided engineering (CAE) tools with the product model and make improvements to the design process continuously.

There are various characteristics associated with an effective concurrent engineering team, including dynamic leadership, positive relationships among members, open and frank communication, good cooperation among members, clearly defined and agreed-upon objectives/goals, existence of healthy and low levels of conflicts, clearly defined process and procedures, and well-defined and accepted functions of team members [12].

From time to time, conflicts do arise between the requirements and desires of each individual member of the concurrent engineering team. The wise leadership can turn such conflicts into healthy dialogues. Some of the guidelines [11, 18] for handling conflicts in the concurrent engineering environment are (1) to place external and internal customers' needs first, (2) to analyze with care each individual team member's inputs, (3) to determine reason for the conflict, (4) to take advantage of compromise as much as possible, and (5) to aim for consensus decisions.

3.2.6 CONCURRENT ENGINEERING RISKS AND BENEFITS EXPERIENCED BY VARIOUS U.S. COMPANIES

The application of the concurrent engineering approach will not automatically produce superior and effective product designs. But there are certain risks associated with the application of the approach. Some of these are as follows [12]:

- A good designer, engineer, etc. may not be a good team player.
- It is more difficult to manage teams than individuals.
- Design for manufacturability (DFM) and design for excellence or all desirable characteristics (DFX) may be practiced without forming a concurrent engineering team.
- Resistance to change can occur during the implementation of the concurrent engineering concept.
- The cost of establishing teams can be quite significant if design, production, and important staff functions are situated at different locations.

There are various companies in the United States that have used the concurrent engineering concept quite successfully, including [11]

- IBM (East Fishkill location): The application of the concurrent engineering concept during the development of an automated electronic design automation system resulted in 50% reduction in product direct labor cost, 65% reduction in processing time for customizing products, 40% reduction in overall digital electronic design cycle, and so on.

- NCR: The application of the concurrent engineering concept during the development of NCR's 2760 electronic cash register reported successes such as 80% reduction in parts and assembly time, 44% reduction in manufacturing costs, 65% reduction in suppliers, 100% reduction in assembly tools, and so on.
- Boeing (Ballistic Systems Division): The use of product development teams, as well as a set of 84 internal improvement initiatives, resulted in 16 to 46% reduction in manufacturing costs, a reduction factor of three in inspection costs, 12% reduction in material shortages, and so on.
- Texas Instruments: The application of the concurrent engineering concept during the redesign of a complex infrared sight resulted in 85% reduction in assembly time, 75% reduction in parts, and 71% reduction in number of steps.

3.3 REVERSE ENGINEERING

Reverse engineering may simply be described as a special case of system engineering that may be applied to rectify defects in or to extend the capabilities of already existing facility. Further, a technologist generally interprets reverse engineering as the task of developing a set of functional specifications for an item/system/equipment other than the original designer based on an analysis of an already existing item/system/equipment [19]. In order to understand a reverse engineering effort clearly, it is useful to understand whether the end objective is to produce a clone or a surrogate item/system/equipment [5]. In the case of producing a clone, it means the exact reproduction of the original (at least as far as circumstances permit). More clearly, the clone reproduction must have the same function, form, operating mechanism, and fit as the original item/system/equipment. In contrast, the surrogate item carries out the same function as the original, in addition to fitting in the same place as the original, but it does not necessarily appear to be the same as the original or use the same operating mechanisms as the original.

Obviously, the reverse engineering function is far more extensive in the case of a clone rather than the surrogate. In addition, the increase in complexity and sophistication of modern equipment has also made the task of producing clones more difficult.

Further discussion on reverse engineering is divided into the following categories [5]:

- fundamental considerations: These include various basic considerations associated with reverse engineering, e.g., specifications, involvement of technical experts, and design factors.

- documentation procedure: This is concerned with recording the information uncovered in the reverse engineering process.
- reverse engineering approach: This is concerned with accomplishing the reverse engineering effort.

Each of these three categories is described below.

3.3.1 FUNDAMENTAL CONSIDERATIONS

There are various basic considerations [5] associated with the reverse engineering effort, including

- item specifications: As in the case of any engineering activity, in the reverse engineering effort, the complex items/products/systems, require two types of specifications: functional and dimensional. The purpose of the functional specifications is to describe the working of the item/product/system, as well as its associated subsystems and their interactions. On the other hand, dimensional specifications include item/parts dimensions, material used in the fabrication of such items/parts, parameter values and their tolerances, and the description of the assembly of those parts during manufacture. Generally, the reverse engineering makes use of the two-step strategy: (1) developing the functional specifications to the level of fully understanding the operating mechanisms, including an action requiring hardware decomposition to a point where the existence of some assemblies is possible, and (2) disassembling the remaining assemblies and isolating all the parts and then measuring such parts to establish their dimensions.
- design factors: It is an accepted fact that the reverse engineering is more cumbersome than executing an original design. Some of the reasons are the inability, in general, to determine the thinking of the designer when the design was developed; determining the crucial parameters with respect to perfomance; determining the treatments applied to the materials; and determining the elements critical to the item operation but that are not essential for the operation of the specific assembly.
- indirect influences: During the reverse engineering effort, these are usually a one-time consideration. A careful consideration to indirect influences prior to undertaking the reverse engineering effort can contribute greatly to the effectiveness of the reverse engineering effort. The indirect influences include manufacturing philosophy, potential product

users, maintenance policy, logistic support philosophy, and the tactical deployment.

- original item specimen: The original specimen of the item to be cloned becomes quite useful to arrive at various kinds of decisions during the reverse engineering process, e.g., testing hypotheses during the reverse engineering process when everything else fails. Therefore, it is absolutely essential to have at least one specimen of the item/product/system to be cloned in its original form.
- technical expert involvement. The reverse engineering effort usually takes input from various technical specialists. Therefore, it is quite crucial to have experts at hand when performing reverse engineering on various elements associated with the item under consideration because there is a better chance that these individuals will be able to identify clues more easily than the less knowledgeable people.

3.3.2 DOCUMENTATION PROCEDURE

Since the findings of the reverse engineering process have to be communicated to others, it is useful to adopt a suitable documentation scheme. Every effort should be made so that the documents used to record uncovered information during the reverse engineering process are competitive with the method employed to guide that process. The documents involved may be divided into three categories [5]:

- equipment breakdown hierarchy or structure document: This provides a mechanism to order the subsystems, their assemblies, their subassemblies, and their elements of a specified system to expedite the specification development for the system under consideration. The equipment breakdown hierarchy document is a critical element of the functional specification where it acts as a vehicle to guide the reverse engineering effort. In addition, the document is the ordering mechanism for developing the configuration document.
- configuration document: The basic purpose of this document is to describe interconnections between various components of a particular item; specifics of the flow of information, energy, or materials between these components; and the function performed by the components. The configuration documentation is composed of several interrelated parts, including functional description(s), block diagram(s), and interface tables.
- performance specification document: This document records the performance specifications for the item undergoing the reverse engineering by

formulating a specification tree having the same structure as the equipment breakdown hierarchy or structure. The specification tree entries, at all levels but the lowest (piece component), provide description of the item's functional aspects. The performance specification entries at the piece component level are basically of dimensional type.

3.3.3 REVERSE ENGINEERING APPROACH

The purpose of the reverse engineering method is to provide a vehicle to direct work through all associated steps to the release of the specifications, as well as to highlight clues to the technical specialist that certain technical aspects of the work under consideration may not be that critical to the cloning effort success. The approach described in this section is based on two assumptions: (1) the item/product/system under consideration can be characterized as a hierarchical structure, and (2) the approach is repeatedly applied to item/product/system until its reduction to piece parts or components. The grand plan for performing a reverse engineering effort is composed of the following [5]:

- "system-engineering" to develop hypotheses based on available data and to highlight the measurement/test needs
- disassembly to the level necessary to verify or modify the hypotheses, as well as to conduct supporting tests
- further "system-engineering" on the basis of all available data and establishing new hypotheses, as well as making preparation for additional measurement and testing
- further disassembly, measurement, and testing to validate hypotheses and to obtain new information
- continuing the process until the level of understanding is adequate to prepare the dimensional specification

The generic reverse engineering process is composed of five sequenced steps: existing data assimilation; element identification; disassembly; test, analysis, and dimension; and documentation. The purpose of existing data assimilation is to facilitate an understanding of the item/product, including its operation within the overall scheme of things. The element identification step is concerned with postulating how the item/product is reduced into its parts to facilitate the defining of the item/product in greater detail.

The disassembly step is concerned with isolating the item/product parts, highlighting the interconnection among the parts, and developing the interfaces between the parts and the world outside the item/product. The purpose of the test, analysis, and dimension step is to develop item/product specification in terms of item parts. The documentation step is concerned with formalizing the specifications of the item/product under consideration.

3.3.4 REVERSE ENGINEERING TEAM AND APPLICATION CANDIDATES

Performing reverse engineering is not a one-man task; it requires a group of specialists such as engineers, estimators, shop personnel, draftsmen, technicians, and production-manufacturing workers [6].

From time to time specialists in areas such as circuit design, vibration analysis, metallurgy, and ceramics are also required during the reverse engineering effort. However, the size of the team may vary from one project to another. Further, there are only a handful of people who form the core reverse engineering team. Nevertheless, from the consistency and experience aspect, it is generally important to keep the same core team members from project to project. Also, the person selected to lead the team should be a generalist with some knowledge in engineering disciplines, such as electrical, mechanical, manufacturing, and electronics. Because this person has to interact with various people, the possession of good managerial qualities is also essential.

A careful consideration is required in selecting candidates for the application of reverse engineering. Usually, good reverse engineering candidates exhibit symptoms such as excessive cost, high failure rate, and high usage. Additional factors that require consideration are patent right, technical data adequacy and availability, support obsolescence, and lack of supply. All in all, the following characteristics are quite useful to identify good application candidates [18]:

- return on investment
- economics
- technical complexity and criticality
- logistics

3.3.5 REVERSE ENGINEERING BENEFITS, STRATEGIES, AND COMPARISON OF TRADITIONAL AND REVERSE ENGINEERING DESIGN PROCESSES

There are several advantages of practicing reverse engineering, including cost reduction, maintaining high-performance manufacturing capability, and an effective stopgap measure for improving system productivity until needed resources are within reach for full modernization [6]. In general, it may be said that reverse engineering is directed at modernizing single-system elements, rather than total systems, for the purpose of maintaining or increasing system productivity.

For the successful execution of the reverse engineering effort, some of the strategies to be practiced are as follows:

- Aim to expect return on investment for prescreened candidates at least 25:1.
- Plan to expect that a good reverse engineering program will take from 2 to 5 years to become self-sufficient.
- Aim to expect return on investment from high-risk projects at least 200:1.
- Plan to expect from a reverse engineering application a minimum of 25% reduction in the item/product unit cost.
- Make only a moderate investment during the initial stage of the reverse engineering program.

There is a significant difference between traditional and reverse engineering design processes. The stages of a traditional design process may vary from four to twenty-five [18]. For example, the stages of a four-stage process are requirement, design idea, prototype and test, and product. In contrast, the stages of the reverse engineering design process are as follows [6]:

- product
- disassembly
- measure and test
- design recovery
- prototype and test
- reverse engineered product

3.4 REENGINEERING

Reengineering is concerned with the examination and alteration of a system or an item under consideration to reconsitute it in a new form and the ultimate implementation of that form. However, from the corporation reengineering perspective [20], reengineering is defined as the fundamental rethinking and radical redesign of business processes to accomplish dramatic improvements in critical, contemporary measures of performance such as service, quality, cost, and speed.

In any case, the main objective of any reengineering project is to produce a better product/function for the same cost or a lower cost product/function of comparable quality to the current product/function. This means reengineering helps to improve an organization's competitiveness by allowing it to deal with dynamic external environments. Reengineering may be approached from either of the following three perspectives:

- reactive: This means organization is in difficulty or crisis and reengineering is one way to correct this situation.
- proactive: This means the organization in question positions itself for the anticipated changes in the future so that under the changed condition it emerges as a market leader.
- interactive: This means organization would like to be up-to-date with current changes as they occur.

There are many improvement approaches that relate to reengineering [21]: total quality managment (TQM), strategic cost management, benchmarks, integrated product development, horizontal management, concurrent engineering, reverse engineering, activity-based costing, strategic information management, domain engineering, supply chain integration, and so on. However, the entity to be reengineered can be either one of the following:

- product
- process
- systems management
- some appropriate combination of the above

3.4.1 PRODUCT REENGINEERING

In this context the term *reengineering* could be interpreted as some kind of reworking or retrofit of an existing product, which could simply be described as maintenance or refurbishment [22]. Further, maintenance can be viewed from different perspectives: interactive or adaptive, reactive or corrective, proactive or perfective. Alternatively, reengineering may also be interpreted as reverse engineering in which case the characteristics of an existing product are highlighted for the purpose of modifying or reusing the product. These notions contain two important reengineering facets: (1) improvement of the delivered system/product with respect to reliability and safety or satisfying system/product evolving user requirements, and (2) improvement in the understanding of the system/product. This interpretation of reengineering is basically product focused; thus, we call it product reengineering.

Hence, product reengineering may be described as the examination, study, and modification of the internal mechanisms or functionality of an already developed item/product for the purpose of reconstituting it in a modern form with modern desirable features (the action such as this is often taken to benefit from newly emerged technologies), but without making a major change to the item/product pupose and the inherent functionality. There are many synonyms for product reengineering: modernization, retrofit, repair, renewal, redevelopment, and so on.

3.4.2 PROCESS REENGINEERING

Reengineering can also be applied at the process level, which is basically concerned with modifications to the existing standards cycles in use in an organization for the purpose of taking advantage of new and emerging technologies or to meet new needs of a customer associated with an item/product/system in question. In a broader sense, the process level reengineering is composed of synthesis or determination of a desirable process associated with producing an item/product by keeping in mind the generic customer requirements, as well as considering the systems-engineering organization's objectives and crucial capabilities. Thus, in a similar manner to the product reengineering, the process reengineering is defined as the examination, study, and modification of the internal mechanisms or functionality of an already developed process or system engineering life cycle for the purpose of reconstituting it in a modern form with modern functional and nonfunctional features (such an action is taken to benefit from newly emerged technologies or to encompass desirable organizational or technological capabilites), but without making any changes in the basic objective of the process being reengineered. According to Reference [21], AT&T Bell Laboratories practiced a similar approach to process reengineering and experienced benefits such as reduction in engineering change orders, shorter development cycles, satisfied customers with respect to their expectations concerning products in question, and reduction in program and product development costs throughout the life cycle.

3.4.3 SYSTEMS MANAGEMENT LEVEL REENGINEERING

Reengineering at the level of systems management is concerned with potential changes in total business or organizational processes, in addition to the systems acquisition process life cycle. Many people [20, 23–26] have discussed the subject of reengineering the corporation or business reengineering. The reengineering definition [20] given earlier (i.e., reengineering is the fundamental rethinking and radical redesign of business processes to accomplish dramatic improvements in critical, contemporary measures of performance such as service, quality, cost, and speed) is the reengineering at the level of systems management. This definition contains four important terms:

- processes
- radical redesign

- fundamental
- dramatic improvements

According to Reference [23], reengineering and revolution are almost synonymous terms, and there are three types of companies that attempt reengineering: (1) those passing through difficult times, (2) those that anticipate difficulties in the pipeline, and (3) those ambitious and seeking to avoid impending difficulties.

Keeping these factors in mind, Reference [21] defines systems management reengineering as the examination, study, and modification of the internal mechanisms or functionality of already developed system managment processes and procedures in a firm for the purpose of reconstituting them in a modern form and with modern features (an action such as this is often taken to benefit from newly emerged organizational competitiveness need), but without making any changes in the basic objective of the organization.

3.4.4 REENGINEERING SUCCESS AND FAILURE FACTORS

The practice of reengineering is not an automatic success. In fact, a careful consideration in its application is required for ultimate success. The following factors are quite useful in enhancing a successful application of reengineering [21]:

- Establish aggressive, but achievable, reengineering performance goals with respect to results.
- Obtain commitment for a desirable proportion of the chief executive officer's (if applicable) time to the reengineering project, particularly during the deployment phase.
- Assign the task of the reengineering project leadership to a very senior executive (if applicable).
- Analyze with care customer needs, organizational realities, market trends, and strategic economic issues prior to the start of the reengineering effort.
- Carry out a pilot study, as well as prototype the reengineering effort, for obtaining results helpful in making refinement to the reengineering process, increasing communications, and building enthusiasm.

On the other hand, some of the important ways reengineering efforts fail are as follows [27]:

- failure to communicate effectively during the implementation phase

- measuring the reengineering plans and activities but overlooking the results
- allocating average performing people to the task of reengineering
- allowing innovative ideas for reengineering to be overlooked because of politics, risk involved, and so on.

3.4.5 PRODUCT REENGINEERING RELATED RISKS

Reference [28] has presented many product reengineering requirements in the form of risks to be effectively managed during the reengineering process, including

- risk type I: This is known as integration risk and is associated with the reengineered product that cannot be integrated effectively with the already existing systems.
- risk type II: This is known as cost risk and is associated with the reengineered product meeting specifications only; it has major cost overruns.
- risk type III: This is known as schedule risk and is associated with schedule delays in order to have the reengineered product meeting all specifications.
- risk type IV: This is known as maintenance improvement risk and is associated with the reengineered product increasing, rather than decreasing, maintenance difficulties.
- risk type V: This is known as human acceptance risk and is associated with the reengineered product being unsuitable for human interactions.
- risk type VI: This is known as process risk and is associated with the reengineered product that might improve situations, but there exists a certain defective organizational process in which it (the product) is to be used.
- risk type VII: This is known as systems management risk and is associated with the engineered product imposing a technological fix on a situation requiring organizational reengineering at the level of systems management.
- risk type VIII: This is known as tool and method availability risk and is associated with the reengineering of the product on the assumption that the required methods and tools will be available.
- risk type IX: This is known as leadership, strategy, and culture risks and is associated with the reengineered product imposing a technological fix on the organizational environment that may not adapt to the level of the product in question.

- risk type X: This is known as application supportability risk and is associated with the reengineered product's failure to support the application it was originally intended to fulfill.

3.4.6 REENGINEERING MANPOWER

Performing the reengineering effort effectively requires a team of competent individuals. This section presents these people from a corporation reengineering aspect. However, the same idea should also be applicable to product reengineering, and others. Thus, the following elements form the reengineering group [20]:

- group leader: He/she is usually a senior executive or chief executive officer (CEO) who authorizes and motivates the overall reengineering effort within the company. This person's primary role is to act as visionary and motivator, as well as to demonstrate leadership through signals, symbols, and systems [20]. It should be noted that most reengineering failures occur because of the lack of strong, aggressive, committed, and knowledgeable leadership.
- process leader/owner: He/she is usually a senior-level manager with responsibility for a certain process on which the reengineering effort is focused. Basically, this individual's function is to ensure that the reengineering is accomplished successfully at the process level. Obviously, this person does not perform the reengineering task but supervises it so that it gets done in an effective manner.
- reengineering team: This is composed of a group of individuals with sole responsibility to reengineer a certain process by performing analysis of the existing one and overseeing its redesign and implementation. In short, these individuals actually reinvent the business. An ideal amount of the time to be spent by these team members on the reengineering effort should be 100% of their total time; however, a minimum commitment must not be below 75%.
- reengineering steering committee: This is a policy-making body composed of senior executives, and its basic responsibility is to develop an overall reengineering strategy for the organization and to keep track of its progress. Usually, the group leader chairs this committee, although the existence of this committee is optional.
- reengineering czar/expert: This person is responsible for developing methods and tools of reengineering within the organization, in addition to accomplishing synergy across the firm's distinct reengineering efforts. Further, the czar/expert serves as the group leader's reengineering chief of staff. Two important functions of this person are enabling and supporting

each process leader/owner and reengineering team and coordinating all active reengineering activities.

3.5 PROBLEMS

(1) Discuss the historical backgrounds of concurrent engineering, reverse engineering, and reengineering.

(2) Describe the important objectives associated with the application of concurrent engineering.

(3) Discuss at least four methodologies/techniques related to the concurrent engineering process.

(4) Describe in detail the following:
- concurrent engineering team charter
- concurrent engineering team plan
- concurrent engineering team leadership and management

(5) What are the fundamental considerations of the reverse engineering ? Discuss at least three such considerations.

(6) Describe in detail the reverse engineering method.

(7) What are the considerations involved in the selection of candidates for the reverse engineering application?

(8) Discuss the following:
- product reengineering
- process reengineering
- systems management level reengineering

(9) Discuss at least eight product reengineering-related risks.

(10) Describe success and failure factors associated with reengineering.

3.6 REFERENCES

1. Belson, D., Concurrent Engineering, in *Handbook of Design, Manufacturing and Automation*, edited by R.C. Dorf and A. Kusiak, John Wiley and Sons, Inc., New York, 1994, pp. 25–34.
2. Carter, D.E., Baker, B.S., *Concurrent Engineering*, Addison-Wesley Publishing Company, Reading, Massachusetts, 1992.
3. Sanchez, J.M., Priest, J.W., Burnell, L.J., Design Decision Analysis and Expert Systems in Concurrent Engineering, in *Handbook of Design, Manufacturing and Automation*, edited by R.C. Dorf and A. Kusiak, John Wiley and Sons, Inc., New York, 1994, pp. 51–63.
4. Rosenblatt, A., Watson, G.F., Concurrent Engineering, *IEEE Spectrum*, July 1991, pp. 22–23.

5. Rekoff, M.G., On Reverse Engineering, *IEEE Transactions on Systems, Man, and Cybernetics*, Vol. 15, 1985, pp. 244–252.
6. Ingle, K.A., *Reverse Engineering*, McGraw-Hill Book Company, New York, 1994.
7. Chikofsky, E.J., Cross, J.H., Reverse Engineering and Design Recovery: A Taxonomy, *IEEE Software*, January 1990, pp. 13–17.
8. Shina, S.G., Editor, *Successful Implementation of Concurrent Engineering Products and Processes*, Van Nostrand Reinhold Company, New York, 1994.
9. Shina, S.G., *Concurrent Engineering and Design for Manufacture of Electronics Products*, Van Nostrand Reinhold Company, New York, 1991.
10. Shina, S.G., New Rules for World-Class Companies, *IEEE Spectrum*, July 1991, pp. 23–26.
11. Turino, J., *Managing Concurrent Engineering*, Van Nostrand Reinhold Company, New York, 1992.
12. Bralla, J.G., *Design for Excellence*, McGraw-Hill, Inc., New York, 1996.
13. *Webster's Encyclopedic Dictionary*, Lexicon Publications, Inc., New York, 1988.
14. Salomone, T.A., *Concurrent Engineering*, Marcel Dekker, Inc., New York, 1995.
15. Askin, R.G., Sodhi, M., Organization of Teams in Concurrent Engineering, in *Handbook of Design, Manufacturing and Automation*, edited by R.C. Dorf and A. Kusiak, John Wiley and Sons, New York, 1994, pp. 85–105.
16. Hall, D., Concurrent Engineering: Defining Terms and Techniques, *IEEE Spectrum*, July 1991, pp. 24–25.
17. Pugh, S., *Total Design Integrated Methods for Integrated Methods for Successful Product Engineering*, Addison-Wesley Publishing Company, Reading, Massachusetts, 1991.
18. Dhillon, B.S., *Engineering Design*, Richard D. Irwin, Inc., Chicago, 1996.
19. Samuelson, P., Reverse Engineering Someone Else's Software: Is It Legal? *IEEE Software*, January 1990, pp. 90–96.
20. Hammer, M., Champy, J., *Reengineering the Corporation*, Harper Collins Publishers, Inc., New York, 1993.
21. Sage, A.P., Systems Engineering and Systems Management for Reengineering, *Journal of the Systems Software*, Vol. 30, 1995, pp. 3–25.
22. Hunt, V.D., *Reengineering: Leveraging the Power of Integrated Product Development*, Oliver Wright Publications, Essex Junction, Vermont, 1993.
23. Hammer, M., Reengineering Work: Don't Automate, Obliterate, *Harvard Business Review*, Vol. 68, 1990, pp. 104–112.
24. Graham, M.A., Lebaron, M.J., *The Horizontal Revolution: Reengineering Your Organization through Teams*, Jossey-Bass Publishers, San Francisco, 1994.
25. Champy, J., *Reengineering Management*, Harper Collins Publishers, Inc., New York, 1995.
26. Ingstrup, O., *Reengineering in the Public Service*, Explorations No. 6, Canadian Centre for Management Development, Ministry of Supply and Services, Ottawa, 1995.
27. Hall, G., Rosenthal, J., Wade, J., How to Make Reengineering Really Work, *Harvard Business Review*, Vol. 71, 1993, pp. 119–131.
28. Arnold, R.S., Common Risks of Reengineering, *IEEE Comp. Soc. Rev. Eng. Newslett.*, April 1992, pp. 1–2.

CHAPTER 4

Design Reliability

4.1 INTRODUCTION

AS the sophistication and complexity of engineering systems have increased, reliability has become an important design parameter. It is probably reasonable to state that reliability is the most important measure of the integrity of a designed item. Reliability is defined as the probability that an item will perform its designed-for functions satisfactorily for the stated period when used according to specified conditions.

The beginning of the reliability field may be traced to the early 1930s when probability concepts were applied to electric power generation problems [1, 2]. However, usually in the published literature, the starting point of the reliability discipline is regarded as World War II, with the development of German V1 and V2 rockets. In fact, it was Robert Lusser who played an instrumental role in their development with respect to reliability [3]. In 1947, an extensive study concerning failures of over 100,000 tubes (i.e., electronic valves) was jointly conducted by Aeronautical Radio, Inc. and Cornell University. As a result of various reliability-related studies performed by its three services – Air Force, Navy, and Army – the United States Department of Defense established an ad hoc committee on reliability [4] in 1950. A year later, General George C. Marshall, the Secretary of Defense, issued a directive to increase reliability of military electronic equipment. In 1952, Assistant Secretary of Defense (research and engineering) transformed the ad hoc committee to a permanent group known as the Advisory Group on the Reliability of Electronic Equipment (AGREE) [5].

In 1954, the first National Symposium on Reliability and Quality Control was held in the United States, and in 1955, the Institute of Electrical and Electronic Engineers (IEEE) established the Reliability and Quality Control Society.

In 1956, a book entitled *Reliability Factors for Ground Electronic Equipment*, edited by K. Henny, was published [6]. Two important documents on reliability appeared in 1957: (1) the AGREE report and (2) the first military reliability specification, MIL-R-25717 (USAF): Reliability Assurance Program for Electronic Equipment, issued by the United States Air Force (USAF). In 1962, the first master's degree program in system reliability engineering was started by the United States Air Force Institute of Technology, Dayton, Ohio, and in 1966, the Society of Reliability Engineers (SRE) was formed in Buffalo, New York.

Since 1966, hundreds of publications on reliability have appeared [7, 8]. Some of the important periodicals on the subject are as follows:

- *IEEE Transactions on Reliability*
- *Microelectronics and Reliability*
- *Reliability Engineering and System Safety*
- *Journal of Quality and Reliability Management*
- *Reliability Review*
- *Quality and Reliability Engineering International Journal*
- *Reliability, Quality and Safety Engineering*

This chapter describes various different aspects of design reliability

4.2 REASONS FOR CONSIDERING RELIABILITY IN PRODUCT DESIGN

In the past, there have been various reasons for considering reliability during the design phase, including product complexity, competition, public demand, and insertion of reliability-related clauses in design specifications. In fact, the increase in the product complexity has been one of the most important factors for the emphasis on reliability during the product/system design phase. Some of the examples of increase in the product complexity with respect to parts alone are as follows [9–11]:

- A Boeing 747 jumbo jet airplane is composed of approximately 4.5 million parts, including fasteners.
- The 1964 Mariner spacecraft contained approximately 138,000 components.
- In 1935 a farm tractor was made up of 1200 critical parts but in 1990 the total tractor critical parts were estimated to be around 2900.

Many studies have revealed that the most effective for-profit contribution is the reliability professionals' involvement with designers. In fact, past experience has shown that if it costs $1 to correct a design shortcoming prior to the initial drafting release, the cost would increase by 10-fold after the final release, 100-fold at the prototype stage, 1000-fold at the preproduction stage, and 10,000-fold at the production stage. Also, various studies have indicated that usually the greatest causes for the product failures are the design-related problems, especially in the case of electronic equipment. For example, a study conducted by the United States Navy concerning electronic equipment provided the following breakdowns for the causes of failure [9]:

- design: 43%. This includes circuit and component deficiencies (11%), inadequate components (10%), circuit misapplication (12%), design weaknesses and unsuitable materials – mechanical (5%), and unsatisfactory parts – mechanical (5%).
- operation and maintenance: 30%. This includes faulty maintenance (8%), abnormal or accidental condition (12%), and manhandling (10%).
- manufacturing: 20%. This includes defective raw materials (2%) and faulty workmanship, inadequate inspection, and process control (18%).
- miscellaneous: 7%. This includes worn out, old age, and so on.

Failures such as listed below also helped to increase the importance of reliability in design:

- 1963: The U.S.S. Thresher nuclear submarine disaster. This submarine exceeded its maximum test depth and imploded, resulting in 129 deaths.
- 1967: Point Pleasant Bridge (West Virginia/Ohio border) disaster. This bridge collapsed basically because of metal fatigue of a critical eye bar, resulting in 46 deaths.
- 1979: DC-10 airliner crash. This commercial aircraft crashed basically because of the use of poor maintenance procedures, resulting in many deaths.
- 1982: Jarvik artificial heart malfunction. A valve malfunctioned 2 days after this mechanical heart's first human transplant, causing the patient to undergo further operations. Prior to its (i.e., artificial heart's) recalling by the Food and Drug Administration (FDA) because of deficiencies in manufacturing quality, training, and so on, it affected the lives of 157 patients.
- 1986: Space Shuttle Challenger debacle. This disaster occurred because of design defects, resulting in all crew members' deaths.
- 1986: Chernobyl nuclear reactor explosion. This disaster in the former Soviet Union has been attributed to design defects, ultimately resulting in 31 deaths.

4.3 RELIABILITY IN THE SYSTEM DESIGN AND DEVELOPMENT PROCESS, PRODUCTION PROCESS, AND SUPPORT PROCESS

The reliability considerations in system/product design and development process are very important because they determine the reliability of the design itself to a large extent. Many reliability-related tasks are performed during this phase, including establishing reliability requirements definition, developing reliability program plan, reliability modeling, reliability allocations, reliability predictions, failure modes effects and criticality analysis, reliability growth testing, subcontractor/supplier monitoring and control, and environmental stress screening [5]. Tasks such as these, if performed properly, lead to one of the most critical contributions to system/product/equipment design and development.

The reliability of the product/system is not only dependent on the design, but also it is a function of the product/system that is manufactured from that design. Thus, a careful performance of various reliability activities during the production process is absolutely necessary to produce a reliability effective product/system. During the production process, the reliability tasks basically address the design and control of the manufacturing process. These tasks include the use of screening methods to identify defective parts and to highlight flaws introduced into the product/system by the production processes, use of Taguchi methods, and the use of the U.S. Air Force's Variability Reduction Process (VRP) [5]. Basically, VRP includes the performance of various activities to lower the likelihood of undesirable variation. These activities include producing a design with reduced variability, improving processes to achieve reduction in the manufacturing variations, and using methods/techniques to monitor variations. All in all, the important factors of reliability during the production process include the control of the production processes, component/part quality, focus on and control of changes to design/processes/parts, and the control of variability.

During the support phase, the main objective is to keep a reliable product/system functional in a reliable manner. This calls for the following steps:

- introducing changes to improve design
- highlighting potential harmful factors, as well as keeping them from degenerating the design under consideration
- providing appropriate and systematic introduction into service

The operational phase is the ultimate stage where many hidden design problems become quite transparent, and changes to the design may be recommended. Some of the problems could be operator and maintainer failures and the damaging of parts/components during maintenance actions. The potential harmful factors that could alter product/system reliability include changes to the product/system operating modes, logistics support factors, operator maintainer actions, and changes to replacement parts or processes. The poor introduction of the product/system into service may decrease its reliability. Thus, systematically designed steps are necessary to introduce the system/product into operation.

4.4 REDUCING A PRODUCT/SYSTEM LIFE CYCLE COST WHILE IMPROVING ITS RELIABILITY/MAINTAINABILITY/AVAILABILITY

The life cycle cost of a product/system is the total cost occurring over its entire life span (i.e., procurement cost plus ownership cost). There are many ways during design to reduce a product/system life cycle cost while at the same time improving its reliability/maintainability/availability [3] such as the following:

- Minimize the number of components in the product during design.
- Reduce design errors by using design reliability checklists.
- Perform failure modes, effects, and criticality analysis (FMECA) to identify redesign, research, and development areas.
- Use improved and more compatible materials.
- Integrate reliability engineering/product assurance program in development, engineering, purchasing, manufacturing, testing, installation, and so on.
- Allocate scientifically required product/system reliability/maintainability/availability down to all associated parts/components.
- Perform reliability design reviews.
- Use manufacturing reliability checklists.
- Avoid the abuse of the product/equipment/system in the field by including mechanisms such as load and speed limiters, warning labels, and so forth.
- Assure the proper and adequately burned-in/broken-in/debugged/stress-screened of parts prior to their shipment and use.

4.5 EARLY, CHANCE, AND WEAR-OUT FAILURES OF PRODUCT

Usually, in determining the reliability of a product during the design stage, it is assumed that three types of failure are associated with that product during its entire life span. In other words, it is assumed that the hazard rate or

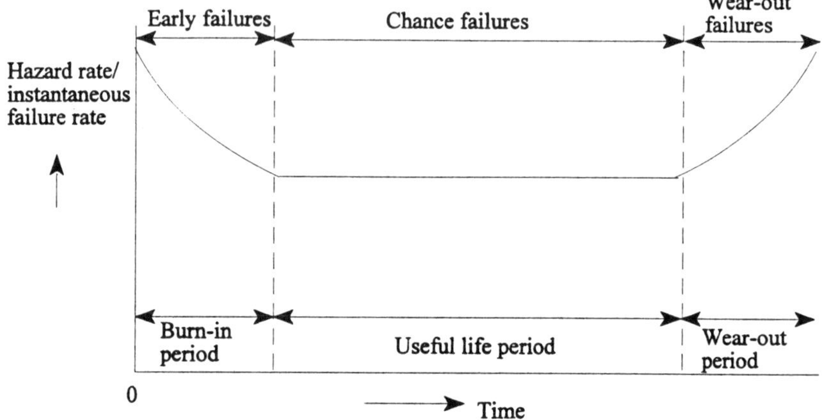

Figure 4.1 Bathtub hazard rate curve.

instantaneous failure rate of a product/item/system follows the shape of a bathtub, as shown in Figure 4.1.

Past experience indicates that many products/items (especially electronic) exhibit bathtub characteristics. Figure 4.1 shows that the bathtub hazard rate curve is divided into three regions: burn-in period, useful life period, and wear-out period. The failures occurring during the burn-in period are called early failures. In fact, during this period the product/item failure rate decreases with time. Some of the causes for the occurrence of the early failures are poor workmanship; use of substandard parts; poor manufacturing processes and techniques; human error; inadequate burn-in; unsatisfactory quality control; faulty installation; failing parts caused by improper storage, packaging, and transportation practices; contamination; inadequate breaking-in; substandard materials; faulty start-up; failed parts replaced with nonscreened parts; and inadequate debugging.

The failures occurring during the useful life period are known as chance, or random, failures. The product/item hazard or failure rate during this period remains constant. Some of the causes for the occurrence of the chance failures are unsatisfactory design in safety factors; abuse; unexplainable causes; overlap of design in strength and actual stress during applications; wrong application; human error; lower expected random strengths; higher expected random loads; unforeseen external events such as earthquakes, tornados, lightening, and floods; defects that escaped detection methods and techniques; and unavoidable failures.

The failures occurring during the wear-out period are called wear-out failures. During this period, the product/item failure rate increases. Some of the causes of the wear-out failures are fatigue, wear, corrosion, poor maintenance practices, short life design, degradation in strength, and aging.

Usually, in performing reliability analysis in the industrial sector, the first (burn-in period) and the last (wear-out period) regions of the bathtub curve are neglected, and only the middle (useful life period) is considered. Thus, the failure rate of the item/product in question during this period is constant, and its times to failures are exponentially distributed. Nevertheless, the entire bathtub hazard rate curve may be represented by the following function for $b = 0.5$ and $\theta = 1$ [10]:

$$\begin{aligned} \lambda(t)\, \alpha\lambda b t^{b-1} &+ (1-\alpha)\theta t^{\theta-1}\gamma e^{\gamma t^{\theta}} \\ \text{for} \quad & \theta, \lambda, \gamma, b > 0 \\ & 0 \le \alpha \le 1 \\ & t \ge 0 \end{aligned} \tag{4.1}$$

where

$\lambda(t)$ = the hazard rate or failure rate at time t
b, θ = shape parameters
γ, λ = scale parameters

4.6 GENERAL FAILURE DENSITY, HAZARD RATE, AND RELIABILITY FUNCTIONS

All three functions are developed as follows [11].

Suppose m identical parts or components are put under test; after some time t, $m_f(t)$, fail and $m_s(t)$ survive. Thus, reliability $R(t)$ is expressed by

$$R(t) = m_s(t)/[m_s(t) + m_f(t)] \tag{4.2}$$

Since the total number of components is $m = m_s(t) + m_f(t)$, Equation (4.2) is rewritten to the following form:

$$R(t) = \frac{m_s(t)}{m} \tag{4.3}$$

The reliability, $R(t)$, at time t of an item/component plus its failure probability, $F(t)$, at time t is expressed by

$$R(t) + F(t) = 1 \tag{4.4}$$

Therefore, we have

$$F(t) = 1 - R(t) \tag{4.5}$$

Substituting Equation (4.3) into Equation (4.5) yields

$$F(t) = 1 - \frac{m_s(t)}{m} \tag{4.6}$$

Since $m = m_s(t) + m_f(t)$, Equation (4.6) yields

$$\begin{aligned} F(t) &= 1 - \left[\frac{m - m_f(t)}{m}\right] \\ &= \frac{m_f(t)}{m} \end{aligned} \tag{4.7}$$

Using Equations (4.4) and (4.7), we get

$$\begin{aligned} R(t) &= 1 - F(t) \\ &= 1 - \frac{m_f(t)}{m} \end{aligned} \tag{4.8}$$

Differentiating Equation (4.8), with respect to time t, yields

$$\frac{dR(t)}{dt} = -\frac{1}{m}\frac{dm_f(t)}{dt} \tag{4.9}$$

In the limiting case, as dt approaches zero, the right-hand side (i.e., without the minus sign) of Equation (4.9) becomes

$$\frac{1}{m}\frac{dm_f(t)}{dt} = f(t) \tag{4.10}$$

where $f(t)$ is the failure density function of an item/component.

Substituting Equation (4.10) into Equation (4.9) yields

$$\frac{dR(t)}{dt} = -f(t) \tag{4.11}$$

Differentiating Equation (4.5) with respect to t yields

$$\frac{dF(t)}{dt} = -\frac{dR(t)}{dt} \tag{4.12}$$

Inserting Equation (4.11) into Equation (4.12) lead to

$$\frac{dF(t)}{dt} = f(t) \tag{4.13}$$

The above expression is the equation for an item's failure density function.

Rearranging Equation (4.9), we get

$$\frac{dm_f(t)}{dt} = -m\frac{dR(t)}{dt} \tag{4.14}$$

Dividing both sides of Equation (4.14) by $m_s(t)$ leads to

$$\frac{1}{m_s(t)}\frac{\mathrm{d}m_f(t)}{\mathrm{d}t} = -\frac{m}{m_s(t)}\frac{\mathrm{d}R(t)}{\mathrm{d}t} \tag{4.15}$$

Alternatively, hazard rate over the interval $[t, t+\Delta t]$ is expressed as the ratio of the number of identical items failing in the time interval to the number of good items at the beginning of the time interval, divided by the length of the time interval [11], i.e.,

$$\lambda(t) = \frac{[m(t) - m(t+\Delta t)]/m(t)}{\Delta t} \tag{4.16}$$

where $m(t)$ is the number of good items at the beginning of the interval.

In the limiting case, as Δt approaches zero, Equation (4.16) is rewritten to the following form:

$$\lambda(t) = -\lim_{\Delta t \to 0} \frac{m(t) - m(t+\Delta t)}{m(t)\,\Delta t} \tag{4.17}$$

Using the $m_f(t)$ and $m_s(t)$ analogy in Equation (4.17), we get

$$\lambda(t) = -\frac{\mathrm{d}m_s(t)}{\mathrm{d}t} \cdot \frac{1}{m_s(t)} \tag{4.18}$$

Since $m_s(t) + m_f(t) = m$, Equation (4.18) becomes

$$\lambda(t) = \frac{\mathrm{d}m_f(t)}{\mathrm{d}t}\frac{1}{m_s(t)} \tag{4.19}$$

It should be noted that the right-hand side of Equation (4.19) and the left-hand side of Equation (4.15) are identical.

Substituting Equation (4.10) into Equation (4.19) yields

$$\lambda(t) = m_f(t)/m_s(t) \tag{4.20}$$

Now, substituting Equation (4.3) into Equation (4.20) leads to

$$\lambda(t) = \frac{f(t)}{R(t)} \tag{4.21}$$

Also, using Equations (4.3) and (4.11) in the right-hand side of Equation (4.15) yields the same result as in the right-hand side of Equation (4.21). Thus, we write

$$\lambda(t) = -\frac{m}{m_s(t)} \cdot \frac{\mathrm{d}R(t)}{\mathrm{d}t} \tag{4.22}$$

Substituting Equation (4.3) into Equation (4.22) leads to

$$\lambda(t) = -\frac{1}{R(t)} \cdot \frac{\mathrm{d}R(t)}{\mathrm{d}t} \tag{4.23}$$

Rearranging Equation (4.23) and then integrating both sides over the time interval $[0, t]$ leads to

$$-\int_0^t \lambda(t)\mathrm{d}t = \int_0^t \frac{1}{R(t)}\mathrm{d}R(t) \tag{4.24}$$

Since at $t = 0$, $R(0) = 1$, Equation (4.24) becomes

$$-\int_0^t \lambda(t)dt = \int_1^{R(t)} \frac{1}{R(t)}\mathrm{d}R(t) \tag{4.25}$$

By integrating the right-hand side of Equation (4.25), we get

$$\ln R(t) = -\int_0^t \lambda(t)\,\mathrm{d}t \tag{4.26}$$

Therefore,

$$R(t) = e^{\int_0^t \lambda(t)\mathrm{d}t} \tag{4.27}$$

The above equation is also known as the general reliability function. It means that Equation (4.27) can be used to obtain a reliability function for any distribution representing times to failure of an item. Some examples of such distributions are Weibull, gamma, exponential, and Rayleigh.

4.6.1 EXAMPLE 4.1

Assume that the times to failure of an electric motor are exponentially distributed as follows:

$$f(t) = \lambda e^{-\lambda t} \tag{4.28}$$

where

$f(t)$ = the failure density function of the motor
λ = the distribution parameter
t = time

Develop expressions for the electric motor reliability and failure probability function and prove that the motor failure rate is constant or is independent of time. By applying the antiderivative theorem to Equation (4.13), we get

$$F(t) = \int_0^t f(t)\,\mathrm{d}t \tag{4.29}$$

Substituting Equation (4.28) into Equation (4.29) results in

$$F(t) = \int_0^t \lambda e^{-\lambda t} \mathrm{d}t = 1 - e^{-\lambda t} \tag{4.30}$$

Thus, the failure probability function of the electric motor is given by Equation (4.30).

Inserting Equation (4.30) into Equation (4.4) and rearranging yields

$$R(t) = 1 - (1 - e^{-\lambda t}) = e^{-\lambda t} \tag{4.31}$$

The above equation is for the reliability function of the electric motor.

Substituting Equations (4.28) and (4.31) into Relationship (4.21), the electric motor's constant failure rate is

$$\lambda(t) = \frac{\lambda e^{-\lambda t}}{e^{-\lambda t}} = \lambda \tag{4.32}$$

Obviously, the right-hand side of Equation (4.32) is independent of time. Thus, the electric motor's failure rate is constant.

4.6.2 EXAMPLE 4.2

Prove by using Equation (4.32) that the reliability expression generated by Equation (4.27) is the same as Equation (4.31). Inserting Equation (4.32) into Equation (4.27) yields

$$R(t) = e^{-\int_0^t \lambda \mathrm{d}t} = e^{-\lambda t} \tag{4.33}$$

It is proven that Equations (4.31) and (4.33) are identical.

4.7 ITEM MEAN TIME TO FAILURE

Mean time to failure (MTTF) is an important design parameter. Many times, in design specifications of products to be developed, its value is defined. For example, the required MTTF of the Canada Arm (i.e., the arm of the Space Shuttle) was 2000 hours. In addition, in any design effort, every effort must be made to maximize the MTTF of the item under consideration within given constraints.

Mathematically, the expected value of the continuous random variable called time to failure is known as the MTTF and is expressed by

$$\text{MTTF} = \int_0^\infty t f(t)\,\mathrm{d}t \tag{4.34}$$

or

$$\text{MTTF} = \int_0^\infty R(t)\,\mathrm{d}t \tag{4.35}$$

or

$$\text{MTTF} = \lim_{s \to 0} R(s) \tag{4.36}$$

where s is the Laplace transform variable.

Often, in reliability analysis, Equation (4.35) is used.

4.7.1 EXAMPLE 4.3

Using Equations (4.28) and (4.31), prove that the electric motor MTTF obtained by applying Equations (4.34), (4.35), and (4.36) is the same. Inserting Equation (4.28) into Equation (4.34) yields

$$\begin{aligned} \text{MTTF} &= \int_0^\infty t\lambda e^{-\lambda t}\,\mathrm{d}t \\ &= [-te^{-\lambda t}]_0^\infty - \left[-\frac{e^{-\lambda t}}{\lambda}\right]_0^\infty \\ &= \frac{1}{\lambda} \end{aligned} \tag{4.37}$$

Substituting Equation (4.31) into Equation (4.35), we get

$$\begin{aligned} \text{MTTF} &= \int_0^\infty e^{-\lambda t}\,\mathrm{d}t \\ &= \left[\frac{e^{-\lambda t}}{-\lambda}\right]_0^\infty \\ &= \frac{1}{\lambda} \end{aligned} \tag{4.38}$$

Taking the Laplace transform of Equation (4.31) results in

$$R(s) = \frac{1}{s + \lambda} \tag{4.39}$$

Using Equation (4.39) in Relationship (4.36) leads to

$$\text{MTTF} = \lim_{s \to 0} \frac{1}{(s+\lambda)} = \frac{1}{\lambda} \tag{4.40}$$

Thus, the MTTF result of Equations (4.37), (4.38), and (4.40) is the same.

4.7.2 EXAMPLE 4.4

Assume that a diesel engine's times to failure are Rayleigh and Weibull distributed with the following reliability functions, respectively:

$$R(t) = e^{-(\beta t^2)/2} \tag{4.41}$$

and

$$R(t) = e^{-(t^a)/\theta} \tag{4.42}$$

where

a = the Weibull distribution shape parameter
θ = the Weibull distribution scale parameter
β = the Rayleigh distribution scale parameter

Obtain expressions for the diesel engine's MTTF using Equations (4.41) and (4.42). Substituting Equations (4.41) and (4.42) into Equation (4.35), respectively, we get

$$\begin{aligned} \text{MTTF} &= \int_0^\infty e^{-(\beta t^2)/2} \text{d}t \\ &= \frac{\Gamma(1/2)}{2(\beta/2)^{1/2}} = (\pi/2\beta)^{1/2} \end{aligned} \tag{4.43}$$

and

$$\text{MTTF} = \int_0^\infty e^{-t^a/\theta} \text{d}t \tag{4.44}$$

If we let $y = t^a/\theta$, Equation (4.44) becomes

$$\text{MTTF} = \frac{\theta^{1/a}}{a} \int_0^\infty e^{-y} y^n \text{d}y \tag{4.45}$$

where

$$n = \left(\frac{1}{a} - 1\right)$$

Intergrating Equation (4.45), we get

$$\text{MTTF} = \theta^{1/a}\,\Gamma\left(1+\frac{1}{a}\right) \tag{4.46}$$

where $\Gamma(\cdot)$ is the gamma function. Thus, the diesel engine's MTTF for Rayleigh and Weibull distributions is given by Equations (4.43) and (4.46), respectively.

4.8 COMMON RELIABILITY NETWORKS

In design work, the product/system components may form various kinds of configurations with respect to reliability (e.g., series, parallel, or standby). Therefore, this section presents reliability analysis of some of the common configurations or networks that may occur during product/system design [12, 13].

4.8.1 SERIES NETWORK

The series network is the simplest of all reliability networks. In this network, it is assumed that all the components act in series. In other words, if any one of the components fails, the network or system fails. It is important to remember that the network components themselves need not be physically or topologically in series, but what is required is that all the components must operate successfully for the system/network to succeed. A block diagram of n components/units series network/system is shown in Figure 4.2. Each block in Figure 4.2 represents a unit/component. In Figure 4.2, if we let y_i denote the event that the ith unit/component is successful, then the series system reliability, R_s, or probability of success is given by

$$R_s = P(y_1 y_2 y_3 y_n) \tag{4.47}$$

where $P(y_1 y_2 y_3 y_n)$ is the probability of occurrence of events y_1, y_2, y_3 and event y_n.

For all independent events or units/components, Equation (4.47) becomes

$$\begin{aligned} R_s &= P(y_1)\,P(y_2)\,P(y_3)\cdots P(y_n) \\ &= \prod_{i=1}^{n} P(y_i) \\ &= \prod_{i=1}^{n} R_i \end{aligned} \tag{4.48}$$

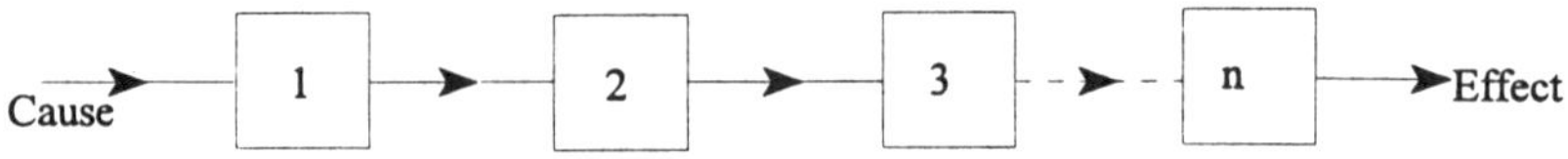

Figure 4.2 A system/network with n units/components in series.

where

$P(y_i)$ = the probability of success of event y_i, for $i = 1, 2, 3, \ldots, n$
R_i = the reliability or probability of success of event y_i or component i, for $i = 1, 2, 3, \ldots, n$

Since the reliability of an item plus its failure probability is always equal to unity, the series network failure probability F_s is expressed by

$$F_s = 1 - \prod_{i=1}^{n} R_i \tag{4.49}$$

or

$$F_s = 1 - \prod_{i=1}^{n} (1 - F_i) \tag{4.50}$$

where F_i is the failure probability of unit/component i, for $i = 1, 2, 3, n$

For constant failure rate of unit/component i, using Equation (4.31), the unit/component i reliability is

$$R_i(t) = e^{-\lambda_i t} \tag{4.51}$$

where λ_i is the unit/component i constant failure rate instead of the electric motor i.

Inserting Equation (4.51) into Equation (4.48), we get

$$R_s(t) = \prod_{i=1}^{n} e^{-\lambda_i t} \tag{4.52}$$

This equation is the expression for the series system/network reliability at time t.

Integrating Equation (4.52) over the time interval $[0, \infty]$ yields the following expression for the series system mean time to failure:

$$\begin{aligned} \text{MTTF}_\text{s} &= \int_0^\infty \left(\prod_{i=1}^{n} e^{-\lambda_i t} \right) \mathrm{d}t \\ &= 1 \Big/ \sum_{i=1}^{n} \lambda_i \end{aligned} \tag{4.53}$$

where MTTF_s is the series system mean time to failure.

Equation (4.53) is often the basis for the specification in the system/product design specification that the system MTTF must be calculated by adding the failure rates of all components and then taking the reciprocal of the resulting value.

Similarly, for Rayleigh and Weibull distributed times to failure, using Equations (4.35), (4.41), and (4.48), the series system MTTF expressions are given by

$$\text{MTTF}_{\text{sr}} = \int_0^\infty \left(\prod_{i=1}^n e^{-\beta_i t^2/2} \right) \mathrm{d}t$$

$$= \left(\pi/2 \sum_{i=1}^n \beta_i \right)^{1/2} \tag{4.54}$$

and

$$\text{MTTF}_{\text{sw}} = \int_0^\infty \left(\prod_{i=1}^n e^{-t^a/\theta_i} \right) \mathrm{d}t$$

$$= \left(\sum_{i=1}^n \theta_i \right)^{1/a} \Gamma\left(1 + \frac{1}{a}\right) \tag{4.55}$$

where

MTTF_{sr} = the series system mean time to failure when the components' failure times are Rayleigh distributed

MTTF_{sw} = the series system mean time to failure when the components' failure times are Weibull distributed

Example 4.4

An aircraft has three identical and independent engines. Each engine's probability of failure is 0.01. Calculate the aircraft's probability of flying successfully with respect to engines, if all the engines are required for the aircraft to fly. Comment on the result.

The aircraft's reliability is

$$R = 1 - 0.01 = 0.99$$

Using the result in Equation (4.47) yields

$$R_s = (0.99)^3 = 0.9703$$

Thus, the aircraft's reliability with respect to engines is 0.9703. And obviously, it is lower than the reliability of the individual engine. In any case, it should be noted that the series network reliability is always lower than its components'/units' reliability.

Example 4.5

Assume that in Example 4.4 the engine failure rate is 0.0002 failures per hour. Calculate the aircraft's reliability with respect to its three engines for a 12-hour flying mission.

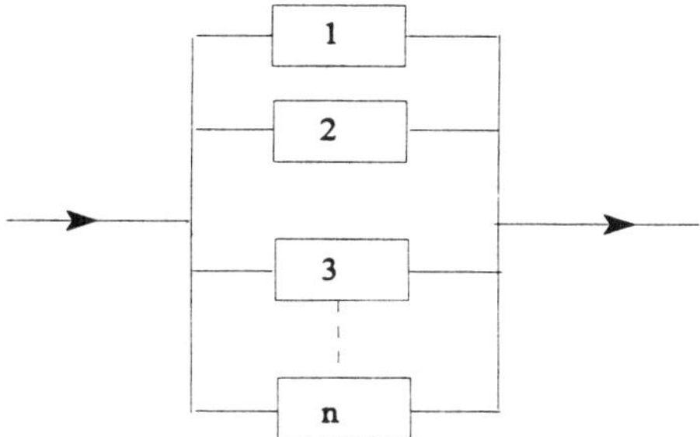

Figure 4.3 Block diagram of an n unit parallel system.

Inserting the given data into Equation (4.52), we get

$$R_s(12) = e^{-(0.0006)(12)}$$
$$= 0.9928$$

Thus, the aircraft's reliability with respect to engines is 0.9928.

4.8.2 PARALLEL NETWORK

This is one form of redundancy used in the industrial sector to increase item/ product reliability. In the case of parallel network, all the units are active, and at least one of these units must work normally for the system success.

A block diagram of n component/unit parallel network/system is shown in Figure 4.3. For n independent units, the failure probability of the parallel network/system shown in Figure 4.3 is

$$\begin{aligned} F_p &= P(\bar{y}_1 \bar{y}_2 \bar{y}_3 \bar{y}_n) \\ &= P(\bar{y}_1) P(\bar{y}_2) P(\bar{y}_3) \cdots P(\bar{y}_n) \\ &= F_1 F_2 F_3 \cdots F_n \end{aligned} \tag{4.56}$$

where

F_p = the failure probability of the parallel system with n units

$\bar{y}_i$ = indication that the ith unit is not successful, for $i = 1, 2, 3, \ldots, n$

$P(\bar{y}_1 \bar{y}_2 \bar{y}_3 \bar{y}_n)$ = the probability that events y_1, y_2, y_3, and y_n are not successful

$P(\bar{y}_i)$ = the failure probability of unit i; for $i = 1, 2, 3, \ldots, n$

Thus, the reliability of the parallel system is

$$\begin{aligned} R_p &= 1 - F_p \\ &= 1 - F_2 F_2 F_3 \cdots F_n \\ &= 1 - (1 - R_1)(1 - R_2)(1 - R_3) \cdots (1 - R_n) \end{aligned} \tag{4.57}$$

where R_p is the reliability of the parallel system.

For constant failure rate of unit/component i, substituting Equation (4.51) into Equation (4.58) yields

$$R_p(t) = 1 - \prod_{i=1}^{n} (1 - e^{-\lambda_i t}) \tag{4.58}$$

where $R_p(t)$ is the parallel network reliability at time t.

Substituting Equation (4.58) into Equation (4.35) and integrating yields the following expression for the parallel system mean time to failure:

$$\begin{aligned} \mathrm{MTTF_p} &= \int_0^{\infty} \left[1 - \prod_{i=1}^{n} \left(1 - e^{-\lambda_i t}\right) \right] \mathrm{d}t \\ &= \sum_{i=1}^{n} \frac{1}{\lambda_i} - \sum_{i=1}^{n} \sum_{\substack{j=1 \\ i \neq j}}^{n} \frac{1}{\lambda_i + \lambda_j} + \sum_{i=1}^{n} \sum_{\substack{j=1 \\ i \neq j \neq k}}^{n} \sum_{k=1}^{n} \left(\frac{1}{\lambda_i + \lambda_j + \lambda_k} \right) - \cdots \\ &\quad + (-1)^{n+1} \frac{1}{\sum_{i=1}^{n} \lambda_i} \end{aligned} \tag{4.59}$$

where $\mathrm{MTTF_p}$ is the parallel system mean time to failure.

For identical units, Equation (4.59) simplifies to

$$\mathrm{MTTF_p} = \frac{1}{\lambda} + \frac{1}{2\lambda} + \frac{1}{3\lambda} + \cdots + \frac{1}{n\lambda} \tag{4.60}$$

Similarly,for Rayleigh and Weibull distributed times to failure, using Equations (4.35), (4.41), (4.42) and (4.57), the parallel system mean time to failure expressions are as follows [14]:

$$\begin{aligned} \mathrm{MTTF_{pr}} &= \sum_{i=1}^{n} \left(\frac{\pi}{2\beta_i} \right)^{1/2} - \sum_{i=1}^{n-1} \sum_{j=i+1}^{n} \left(\frac{\pi}{2(\beta_i + \beta_j)} \right)^{1/2} \\ &\quad + \sum_{i=1}^{n-2} \sum_{j=i+1}^{n-1} \sum_{k=j+1}^{n} \left(\frac{\pi}{2(\beta_i + \beta_j + \beta_k)} \right)^{1/2} - \cdots \end{aligned} \tag{4.61}$$

and

$$\text{MTTF}_{\text{pw}} = \Gamma\left(1+\frac{1}{a}\right)\left[\sum_{i=1}^{n}\theta_i^{1/a} - \sum_{i=1}^{n-1}\sum_{j=i+1}^{n}(\theta_i+\theta_j)^{1/a} + \cdots + (-1)^{n+1}\left\{\sum_{i=1}^{n}\theta_i\right\}^{1/a}\right] \quad (4.62)$$

where

MTTF_{pr} = the parallel system mean time to failure when the components' failure times are Rayleigh distributed

MTTF_{pw} = the parallel system mean time to failure when the components' failure times are Weibull distributed

Example 4.6

A passenger aircraft has three active, independent, and identical engines. The probability of failure of each engine is 0.1. Calculate the probability of success of the aircraft flying with respect to engines if at least one engine must operate normally for the aircraft to fly successfully.

Substituting the specified data for the engine failure probability into Equation (4.56), the failure probability of all three engines failing is

$$\begin{aligned} F_p &= (0.1)(0.1)(0.1) \\ &= 0.001 \end{aligned}$$

Subtracting the above result from unity, the probability of success of the aircraft flying with respect to engines is

$$\begin{aligned} R &= 1 - 0.001 \\ &= 0.999 \end{aligned}$$

Example 4.7

Assume that two independent and identical mechanical pumps are operating in parallel. In other words, at least one of the pumps must operate normally for the system success. Each pump's failure rate is 0.0004 failures per hour. Calculate the pump system mean time to failure and the improvement in mean time to failure by adding a redundant pump in parallel.

Inserting the given data into Equation (4.60), we get

$$\begin{aligned} \text{MTTF} &= \frac{1}{\lambda} + \frac{1}{2\lambda} = \frac{3}{2\lambda} \\ &= \frac{1.5}{0.0004} = 3750 \text{ hours} \end{aligned}$$

The mean time to failure of the single pump is

$$\text{MTTF} = \frac{1}{\lambda} = \frac{1}{0.0004} = 2500 \text{ hours}$$

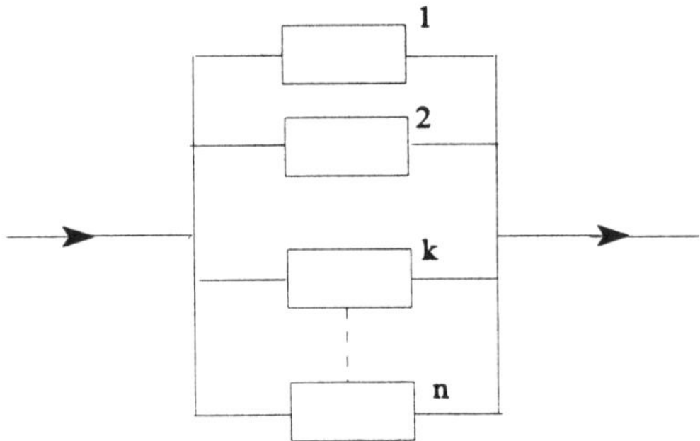

Figure 4.4 Block diagram of a k-out-of-n unit system.

The difference between the above two results is

$$\begin{aligned}\Delta &= 3750 - 2500\\ &= 1250 \text{ hours}\end{aligned}$$

The redundant pump has helped to improve system mean time to failure by 1250 hours, in other words from 2500 hours to 3750 hours.

4.8.3 K-OUT-OF-N UNIT NETWORK

Sometimes, this is also called partially redundant network. In this case, a total of n units are active, and at least k units must operate normally for the system success. Figure 4.4 shows the block diagram of an n component/unit k-out-of-n unit system.

It should be noted that series and parallel networks or configurations are the special cases of this network: $k = n$ (series network), and $k = 1$ (parallel network). For independent and identical n units, the k-out-of-n unit system reliability is

$$R_{k/n} = \sum_{i=k}^{n} \binom{n}{i} R^i (1 - R)^{n-i} \tag{4.63}$$

where R is the unit/component reliability.

$$\binom{n}{i} \equiv \frac{n!}{i!(n-i)!} \tag{4.64}$$

For $k = n$, Equation (4.63) becomes

$$\begin{aligned}R_{n/n} &= \sum_{i=n}^{n} \binom{n}{i} R^i (1 - R)^{n-i}\\ &= R^n \end{aligned} \tag{4.65}$$

The above equation is for the series system with identical units' reliability.

Similarly, for $k = 1$, Equation (4.63) becomes

$$R_{1/n} = \sum_{i=1}^{n} \binom{n}{i} R^i (1 - R)^{n-i}$$

$$R_{1/n} = 1 - (1 - R)^n \tag{4.66}$$

Equation (4.66) is for the parallel system with identical units' reliability.

For a constant failure rate of a unit/component, substituting Equation (4.31) for a unit reliability, instead of the electric motor reliability, into Equation (4.63), we get

$$R_{k/n}(t) = \sum_{i=k}^{n} \binom{n}{i} e^{-i\lambda t} (1 - e^{-\lambda t})^{n-i} \tag{4.67}$$

Inserting Equation (4.35) into Equation (4.67) leads to

$$\begin{aligned} \text{MTTF}_{k/n} &= \int_0^\infty \sum_{i=k}^{n} \binom{n}{i} e^{-i\lambda t} (1 - e^{-\lambda t})^{n-i} \mathrm{d}t \\ &= \frac{1}{\lambda} \sum_{i=k}^{n} \frac{1}{i} \end{aligned} \tag{4.68}$$

where $\text{MTTF}_{k/n}$ is the k-out-of-n unit system mean time to failure.

Example 4.8

An aircraft has three independent and identical engines operating simultaneously. Each engine's reliability is 0.92. Calculate the probability of the aircraft flying if at least two engines must function successfully for the aircraft to fly normally.

Substituting the given data into Equation (4.63) results in

$$\begin{aligned} R_{2/3} &= \sum_{i=2}^{3} \binom{3}{i} R^i (1 - R)^{3-i} \\ &= 3R^2 - 2R^3 \end{aligned} \tag{4.69}$$

Inserting the given engine reliability data into Equation (4.69) yields

$$\begin{aligned} R_{2/3} &= 3(0.92)^2 - 2(0.92)^3 \\ &= 0.9818 \end{aligned}$$

Thus, the probability of success of the aircraft flying is 0.9818.

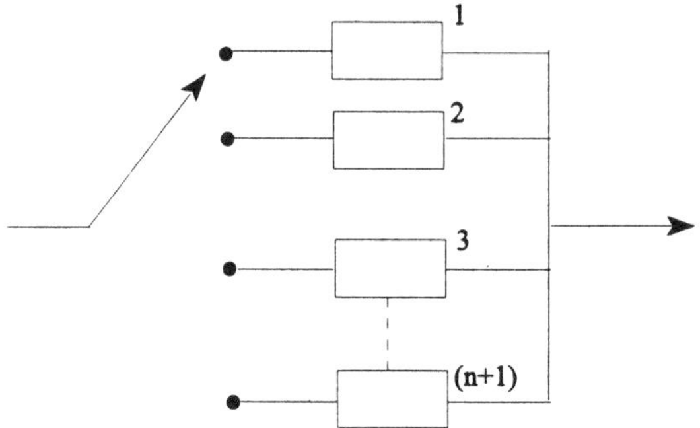

Figure 4.5 The block diagram of an $(n + 1)$ unit standby system.

4.8.4 STANDBY SYSTEM

This is another form of a redundant configuration in which one unit is operating and the remaining n units are on standby. As soon as the operating unit fails, it is immediately replaced with one of the standby units. The block diagram for the configuration is shown in Figure 4.5.

For independent and identical $(n + 1)$ units, the standby system reliability is

$$R_s(t) = \sum_{i=0}^{n} \left\{ \int_0^t \lambda(t)\mathrm{d}t \right\}^i \left\{ \exp\left(- \int_0^t \lambda(t)\mathrm{d}t \right) \right\} \Big/ i! \tag{4.70}$$

where

$R_s(t)$ = the standby system reliability at time t
$\lambda(t)$ = the time dependent failure rate of the unit

The above equation is subject to the following two additional assumptions:

- The switching mechanism is perfect.
- The standby units are as good as new.

For constant failure rate of a unit, i.e., $\lambda(t) = \lambda$, Equation (4.70) simplifies to

$$\begin{aligned} R_s(t) &= \sum_{i=0}^{n} \left\{ \int_0^t \lambda \mathrm{d}t \right\}^i \left\{ \exp\left(- \int_0^t \lambda \mathrm{d}t \right) \right\} \Big/ i! \\ &= \sum_{i=0}^{n} \{\lambda t\}^i e^{-\lambda t} \Big/ i! \end{aligned} \tag{4.71}$$

Inserting Equation (4.71) into Equation (4.35), we get the following expression for the standby system mean time to failure [15]:

$$\begin{aligned}\mathrm{MTTF_{ss}} &= \int_0^\infty \left[\sum_{i=0}^{n} (\lambda t)^i e^{-\lambda t} \Big/ i!\right] dt \\ &= \sum_{i=0}^{n} \left(\frac{\lambda^i}{i!}\right) \int_0^\infty t^i e^{-\lambda t} dt \\ &= \sum_{i=0}^{n} \left(\frac{\lambda^i}{i!}\right) \frac{\Gamma(i+1)}{\lambda^{i+1}} \\ &= \frac{n+1}{\lambda} \end{aligned} \tag{4.72}$$

where $\mathrm{MTTF_{ss}}$ is the standby system mean time to failure.

Example 4.9

A system used in a nuclear reactor has been triplicated because of its criticality. However, only one of the triplicated units is used at a time, and the remaining two units are kept as standby. The failure rate of the single unit is 0.0001 failures/hour. Calculate the standby system reliability for a 200-hour mission if the standby unit switching in mechanism is perfect or never fails and the triplicated units are identical and independent.

For specified data from Equation (4.71), we get

$$\begin{aligned} R_s(200) &= \sum_{i=0}^{2} [(0.0001)(200)]^i e^{-(0.0001)(200)} / i! \\ &= \sum_{i=0}^{2} (0.02)^i (0.9802)/i! \\ &= \frac{0.9802}{0!} + \frac{(.02)(0.9802)}{1!} + \frac{(0.02)^2\,(0.9802)}{2!} \\ &= 0.9999 \end{aligned}$$

Thus, the reliability of the triplicated system is 0.9999.

4.9 RELIABILITY ASSESSMENT METHODS

Ever since the inception of the reliability field, a large number of reliability evaluation methods have been developed. As in the case of any other method, each of these methods has its strong and weak points. These methods are particularly useful to assess reliability of systems more complex than the standard reliability networks. In addition, the sound judgement of design professionals is required in selecting these methods for an application because a certain method could be more effective than others under specific conditions.

This section presents some of these methods particularly useful for design work [12, 13, 15–17].

4.9.1 FAILURE MODES AND EFFECTS ANALYSIS

Failure mode and effects analysis (FMEA) is basically a qualitative approach and is the most widely used method in the design of engineering systems. Also, it can be used to analyze nonengineering systems. The history of FMEA dates back to the early 1950s when it was used to assess the designs of flight control systems [18].

When the criticality analysis (CA) is added to the FMEA process, the method is referred to as failure modes, effects, and criticality analysis (FMECA). Nevertheless, FMEA basically involves the listing of all parts/components in a system under design and identifying each component's failure modes and their (failure modes') anticipated effects. Reference [19] provides a comprehensive list of publications on FMEA/FMECA.

FMEA usually begins during the early stages of design, and the following seven steps are involved:

- Define boundaries and detailed requirements of the product/system under design.
- List all components/subsystems associated with the system/product.
- Identify all possible failure modes of each component and other related information.
- Assign failure rate/probability to each component failure mode.
- Determine each failure mode's effects on surrounding parts/subsystems/ plant/ humans.
- Enter remarks for each identified failure mode.
- Review each crucial failure mode and initiate appropriate measures.

During FMEA performance, various types of questions are asked regarding each and every system component, including (1) What are the possible failure modes of the component? (2) What are the consequences/effects of the each failure mode? (3) How critical are the consequences/effects? (4) How is failure detected? and (5) What are the possible safeguards against the failure in question? In addition, it should be remembered that the types of questions asked depend on the scope and purpose of the analysis under consideration.

There are many uses of FMEA, including identifying weak spots in design; ensuring the understanding of all possible failure modes and their anticipated effects; choosing design alternatives during the early stages of design; serving as a basis for design improvement actions, establishing corrective action priorities, and recommending test programs; and providing assistance in troubleshooting of operational problems in existing systems.

One important limitation of the FMEA is that it is a "single-failure analysis approach." In other words, it considers each failure mode individually. More specifically, FMEA is not suitable for evaluating the combined effects of two or more failures.

4.9.2 FAULT TREE ANALYSIS

The fault tree analysis (FTA) approach is one of the most widely used methods to analyze engineering designs with respect to their reliability in the industrial sector, particularly in nuclear power generation area. In comparison to FMEA, FTA is event-oriented, as opposed to the failure orientation of the FMEA method.

The FTA approach was originated at Bell Telephone Laboratories in the early 1960s to analyze the Minuteman Launch Control System with respect to reliability. Since those years, a vast amount of literature on the subject has been published [7]. Two particular documents describing this method in detail are References [12, 20].

The FTA method makes use of a large number of symbols [12, 21]. The four basic symbols are presented in Figure 4.6(a)–(d). The AND gate symbol denotes that an output fault event only occurs if all the input fault events occur.

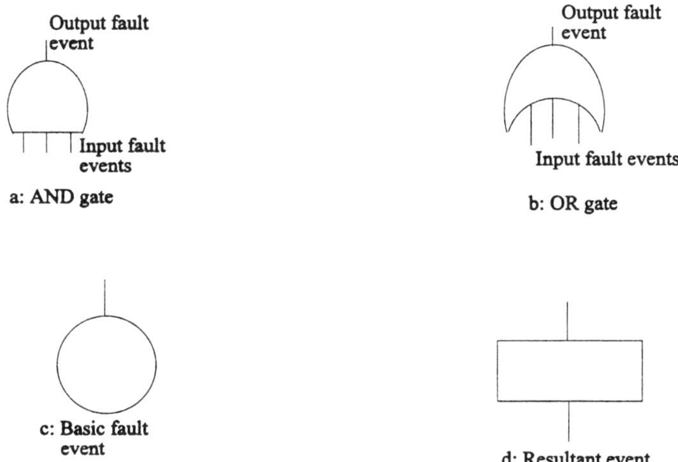

Figure 4.6 Basic fault tree symbols.

On the other hand, the OR gate symbol denotes that an output fault event occurs if one or more of the input fault events occur. The circle represents a basic fault event or the failure of an elementary component. Usually, it is assigned data such as the occurrence probability, failure rate, and repair rate. The rectangle represents a resultant event that results from the combination of fault events through the input of a logic gate.

The following basic steps are involved in developing a fault tree:

- Establish system definition and understand the system thoroughly.
- Construct the fault tree using logic and other symbols.
- Evaluate the fault tree qualitatively.
- Collect appropriate basic data (e.g., elementary components' failure rates, repair rates, and failure occurrence probability).
- Evaluate the fault tree quantitatively.
- Recommend corrective measures.

In particular, it should be noted that the fault construction begins from the undesirable event called the top event and then successively asking the question "How could this event occur?" until reaching the desirable basic fault events.

Probability Evaluation of Fault Trees

Under various conditions, it may be desirable to predict the failure probability of the system/product under design. Before this could be achieved by using the fault tree concept, the estimation of the probability of occurrence of the logic gates' output fault event is essential. Thus, the probability of occurrence of an OR gate's output fault event is expressed by [12]

$$P(E_0) = 1 - \prod_{i=1}^{n}(1 - P(E_i)) \tag{4.73}$$

where

$P(E_0)$ = the probability of occurrence of OR gate's output fault event E_0
n = the number of input fault events
$P(E_i)$ = the probability of occurrence of input fault event E_i; for $i = 1, 2, 3, \ldots, n$

Similarly, the probability of occurrence of an AND gate's output fault event is expressed by

$$P(E_{0a}) = \prod_{i=1}^{n} P(E_i) \tag{4.74}$$

where $P(E_{0a})$ is the probability of occurrence of AND gate's output fault event E_{0a}.

The FTA concept has various advantages and disadvantages. Its advantages include providing insight into the system/product behavior, ferreting out failures deductively, handling complex systems more easily, providing a visibility tool to involve design professionals to justify design changes and trade-off studies, and providing options to conduct qualitative or quantitative reliability analysis. On the other hand, the FTA concept also has certain disadvantages, including difficulty to check results, being a costly and time consuming method, components with degradation states being difficult to handle, and the method requiring a considerable effort to include all types of common-cause failures.

Example 4.10

Assume that a windowless room has two light bulbs controlled by a single switch. The switch can only fail to close. Construct a fault tree for an undesirable event: a dark room. Calculate the probability of occurrence of having no light in the room if all the basic independent fault events' probability of occurrence is 0.1.

Using Figure 4.6 symbols, the fault tree shown in Figure 4.7 was developed. Figure 4.8 presents the final result for the probability of occurrence of having no light in the room (i.e., top event: dark room).

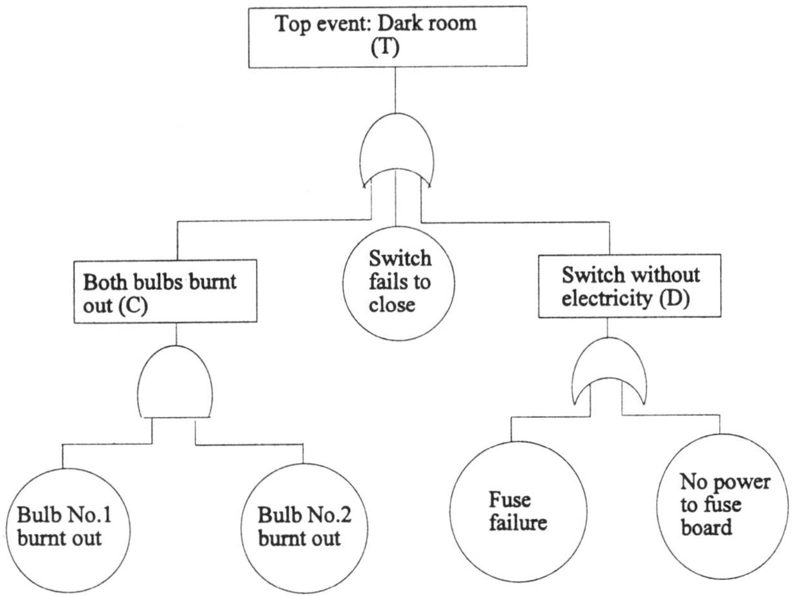

Figure 4.7 Fault tree for having no light in the windowless room.

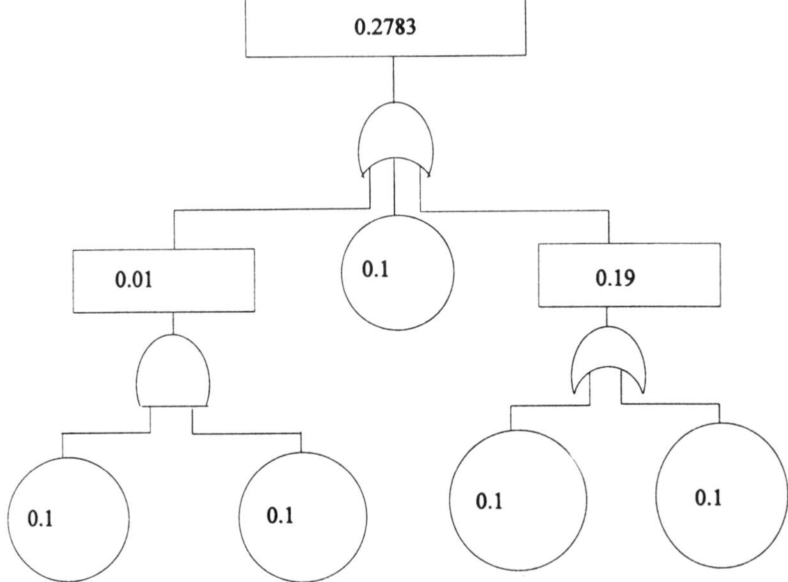

Figure 4.8 Fault tree of Figure 4.7 with fault event occurrence probabilities.

Substituting the given data into Equation (4.74), we obtain the probability of both bulbs burning out:

$$\begin{aligned} P(E_{0a}) = P(C) &= P(E_1)P(E_2) \\ &= (0.1)(0.1) \\ &= 0.01 \end{aligned}$$

Inserting the specified data into Equation (4.73), we get the probability of having a switch without electricity:

$$\begin{aligned} P(E_0) = P(D) &= 1 - (1 - P(E_1))(1 - P(E_2)) \\ &= 1 - (0.9)(0.9) \\ &= 0.19 \end{aligned}$$

Substituting both resulting figures and the given data into Equation (4.73), the probability of having no light in the room is

$$\begin{aligned} P(E_0) = P(T) &= 1 - (1 - P(E_1))(1 - P(E_2))(1 - P(E_3)) \\ &= 1 - (0.99)(0.9)(0.81) \\ &= 0.2783 \end{aligned}$$

Thus, the probability of occurrence of having a dark room is 0.2783.

4.9.3 MARKOV METHOD

This is a powerful reliability evaluation technique and can generally handle more cases than any other method. Its important feature is to handle reliability and availability analyses of repairable systems. The technique can also be used when the components are independent, as well as for systems involving dependent failure and repair modes. The method proceeds by the enumeration of system states, and then the resulting differential equations are solved to obtain various reliability measures. Probably the only serious problem, particularly when constant failure/repair rates are assumed with this method, is that for systems under consideration as the system states increase, it could become unmanageable.

The method is based on the following assumptions [13]:

- All transition rates (e.g., failure and repair rates) associated with the system under consideration are constant.
- The transitional probability from one system state to another in the finite time interval Δt is given by $\alpha \Delta t$, where the α is the constant transition rate (e.g., failure or repair rate) from one system state to another.
- All occurrences are independent of each other.
- The probability of more than one transition in finite time interval Δt from one system state to another is negligible, i.e., $(\alpha \Delta t)(\alpha \Delta t) \rightarrow 0$.

Example 4.11

The failure rate, λ, of a photocopying machine is 0.0002 failures/hour and its repair rate, μ, is 0.0005 repairs/hour. Calculate the photocopying machine availability for a 10-hour mission and its steady-state unavailability by using the Markov method. The photocopying machine state space diagram is shown in Figure 4.9.

Using the Markov method, we get the following equations for state 0 and state 1, respectively, of Figure 4.9.

$$P_1(t + \Delta t) = P_1(t)(1 - \lambda \Delta t) + P_2(t)\mu \Delta t \tag{4.75}$$

$$P_2(t + \Delta t) = P_2(t)(1 - \mu \Delta t) + P_1(t)\lambda \Delta t \tag{4.76}$$

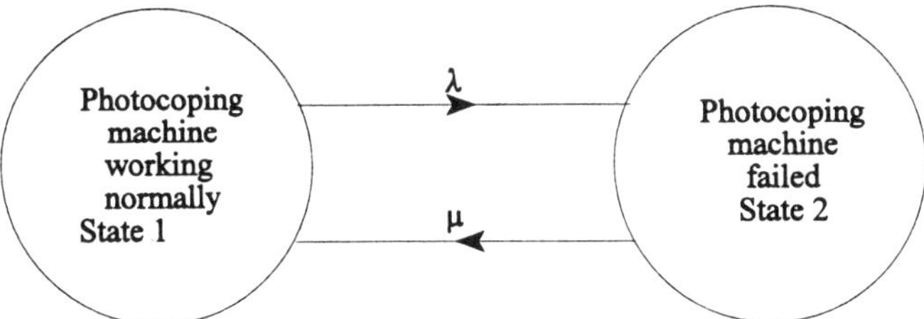

Figure 4.9 State space diagram of the photocopying machine.

where

Δt = the finite time interval
t = is time
λ = the constant failure rate of the photocopying machine
μ = the constant repair rate of the photocopying machine
$\lambda \Delta t$ = the probability of photocopying machine failure in time interval Δt
$\mu \Delta t$ = the probability of photocopying machine repair in time interval Δt
$(1 - \lambda \Delta t)$ = the probability of no failure in time interval Δt
$(1 - \mu \Delta t)$ = the probability of no repair in time interval Δt
$P_i(t)$ = the probability that the photocopying machine is in state i at time t, for $i = 1, 2$
$P_1(t + \Delta t)$ = the probability of the photocopying machine being in working state 1 at time $(t + \Delta t)$
$P_2(t + \Delta t)$ = the probability of the photocopying machine being in failed state 2 at time $(t + \Delta t)$

In the limiting case, Equations (4.75) and (4.76) become

$$\lim_{\Delta t \to 0} \frac{P_1(t + \Delta t) - P_1(t)}{\Delta t} = \frac{\mathrm{d}P_1(t)}{\mathrm{d}t} = P_2(t)\mu - P_1(t)\lambda \tag{4.77}$$

$$\lim_{\Delta t \to 0} \frac{P_2(t + \Delta t) - P_2(t)}{\Delta t} = \frac{\mathrm{d}P_2(t)}{\mathrm{d}t} = P_1(t)\lambda - P_2(t)\mu \tag{4.78}$$

At time $t = 0$, $P_1(0) = 1$, and $P_2(0) = 0$

Solving Equations (4.77) and (4.78) after taking Laplace transforms, we get

$$P_1(s) = (s + \mu)/(s^2 + (\lambda + \mu)s) \tag{4.79}$$

and

$$P_2(s) = \lambda/(s^2 + (\lambda + \mu)s) \tag{4.80}$$

where s is the Laplace transform variable.

Taking the inverse Laplace transforms of Equations (4.79) and (4.80) yields

$$P_1(t) = \frac{\mu}{\lambda + \mu} + \frac{\lambda}{\lambda + \mu} e^{-(\lambda + \mu)t} \tag{4.81}$$

and

$$P_2(t) = \frac{\lambda}{\lambda + \mu} - \frac{\mu}{\lambda + \mu} e^{-(\lambda + \mu)t} \tag{4.82}$$

Equation (4.81) gives the photocopying machine's availability at time t and Equation (4.82) its unavailability at time t. As t becomes very large,

Equations (4.81) and (4.82) reduce to

$$P_1 = \frac{\mu}{\lambda + \mu} \tag{4.83}$$

and

$$P_2 = \frac{\lambda}{\lambda + \mu} \tag{4.84}$$

where

P_1 = the photocopying machine steady-state availability
P_2 = the photocopying machine steady-state unavailability

Since $\lambda = 1/\text{MTTF}$ and $\mu = 1/\text{MTTR}$, Equations (4.83) and (4.84) become

$$P_1 = (\text{MTTF})/(\text{MTTR} + \text{MTTF}) \tag{4.85}$$

$$P_2 = (\text{MTTR})/(\text{MTTR} + \text{MTTF}) \tag{4.86}$$

where

MTTF = the photocopying machine mean time to failure
MTTR = the photocopying machine mean time to repair

Inserting specified data into Equations (4.81) and (4.84) leads to

$$\begin{aligned} P_1(10) &= \frac{0.0005}{0.0002 + 0.0005} + \frac{0.0002}{(0.0002 + 0.0005)} e^{-(0.0002+0.0005)(10)} \\ &= 0.9980 \\ P_2 &= \frac{0.0002}{0.0002 + 0.0005} \\ &= 0.2857 \end{aligned}$$

Thus, the photocopying machine availability is 0.9980 and its steady-state unavailability, 0.2857.

4.9.4 PARTS COUNT METHOD

The parts count method is used to predict the failure rate of a system/product during early design stages and is also quite useful during bid proposal. The method is described in detail in MIL-HDBK-217 [22]. The total product/equipment/system failure rate is expressed by

$$\lambda_T = \sum_{i=1}^{m} N_i (\lambda_G Q_G)_i \text{ failures}/10^6\text{hr} \tag{4.87}$$

where

λ_T = the total failure rate of product/equipment/system
m = the number of different generic component classifications
λ_G = the generic failure rate of the ith generic component expressed in failures/10^6 hr
Q_G = the ith generic component quality factor
N_i = the ith generic component quantity

It should be noted that Equation (4.87) assumes that the entire product/equipment/system will be used under the same environment. The tabulated values for λ_G and Q_G are given in Reference [22].

4.10 RELIABILITY/MAINTAINABILITY STANDARD DOCUMENTS AND FAILURE RATES OF SELECTED ITEMS

Since the 1950s, many different organizations around the world have developed various kinds of standard reliability/maintainability documents to be used during various phases of the product/equipment/system development. In particular, the standard documents prepared by the United States Department of Defense are probably the most widely used in the industrial sector, irrespective of the product/equipment/system under design. Some of those documents are listed below:

- MIL-STD-721, *Definitions of Terms for Reliability and Maintainability*, Department of Defense, Washington, D.C.
- MIL-HDBK-217, *Reliability Prediction of Electronic Equipment*, Department of Defense, Washington, D.C.
- MIL-STD-781D, *Reliability Engineering Development Qualification, and Production*, Department of Defense, Washington, D.C.
- MIL-HDBK-189, *Reliability Growth Management*, Department of Defense, Washington, D.C.
- MIL-STD-785B, *Reliability Program for Systems and Equipment Development and Production*, Department of Defense, Washington, D.C.
- MIL-STD-1629A, *Procedures for Performing a Failure Mode Effects and Criticality Analysis*, Department of Defense, Washington, D.C.
- MIL-STD-470A, *Maintainability Program for Systems and Equipment*, Department of Defense, Washington, D.C.
- MIL-STD-2084 (AS), *General Requirements for Maintainability*, Department of Defense, Washington, D.C.
- MIL-STD-471A, *Maintainability Demonstration*, Department of Defense, Washington, D.C.

- MIL-STD-472, *Maintainability Prediction*, Department of Defense, Washington, D.C.
- MIL-STD-756B, *Reliability Modelling and Prediction*, Department of Defense, Washington, D.C.
- MIL-STD-2155 (AS), *Failure Reporting Analysis and Corrective Action System*, Department of Defense, Washington, D.C.
- MIL-HDBK-338 (Vols. 1 and 2), *Electronic Reliability Design*, Department of Defense, Washington, D.C.
- MIL-STD-790D, *Reliability Assurance Program for Electronic Part Specifications*, Department of Defense, Washington, D.C.

Ever since the inception of the reliability discipline, involved professionals have been collecting various kinds of data. This effort has led to the existence of a large number of data banks, including [16]

- NERC Data: The National Electric Reliability Council (NERC), New York, New York, publishes failure data collected from the United States power plants annually.
- NPRDS Data: On behalf of the U.S. Nuclear Regulatory Commission and the nuclear power industry, the Nuclear Plant Reliability Data System (NPRDS) is managed by the South West Research Institute, San Antonio, Texas.
- FARADA (FAilure RAte DAta): These are in the form of reports containing part failure probabilities derived from the Department of Defense and the National Aeronautics and Space Administration (NASA) experience. The reports are released by the Fleet Missile Systems Analysis and Evaluation Group, Department of Defense, Corona, California.
- NPRD Data: These are in the form of reports containing failure probabilities of nonelectronic parts used by the military. The Nonelectronic Parts Reliability Data (NPRD) reports are released periodically by the Rome Air Development Center, Department of Defense, Griffis Air Force Base, New York.
- GIDEP Data: The Government Industry Data Exchange Program (GIDEP) is a computerized data bank. Originally, GIDEP was jointly established by various agencies: NASA, the Canadian Military Electronics Standards Agency, U.S. Air Force Logistics Command, U.S. Navy, U.S. Army, and U.S. Air Force Systems Command. The data bank is managed by the GIDEP Operations Center, Fleet Missile Systems, Analysis and Evaluation Group, Department of Defense, Corona, California.

Over 350 references concerning failure data are listed in Reference [23].

4.11 PROBLEMS

(1) Write an essay on the historical developments in the reliability field.

(2) What are the principal reasons for considering the reliability factor at the product design stage?

(3) Discuss the reliability-related steps usually taken during the equipment production phase.

(4) What are the possible reasons for having failures during the product's useful life period?

(5) Prove that the mean time to failure (MTTF) of an item is expressed by

$$\text{MTTF} = \lim_{s \to 0} R(s) \tag{4.88}$$

where

s = the Laplace transform variable
$R(s)$ = the Laplace transform of the item reliability

(6) Prove that the mean time to failure of a system (MTTF_s) is given by

$$\text{MTTF}_s = \text{MTTF}_u \sum_{i=1}^{n} \frac{1}{i} \tag{4.89}$$

where

MTTF_u = the mean time to failure of a unit with exponentially distributed failure times
n = the total number of units in the system

State any assumption associated with your derivations.

(7) A system has four independent and identical active units. At least two of the four units must function normally for the system's successful operation. Develop expressions for the system reliability and mean time to failure if its unit failure times are exponentially distributed.

(8) Describe the following:
- standby system
- AND gate
- infant mortality region of the bathtub hazard rate curve

(9) Prove that the mean time to failure of a system is given by

$$\text{MTTF} = \frac{1}{\lambda_1} + \frac{1}{\lambda_2} - \frac{1}{\lambda_1 + \lambda_2} \tag{4.90}$$

where

λ_1 = the unit 1 constant failure rate
λ_2 = the unit 2 constant failure rate

Write down any assumption associated with your proof.

(10) Make comparisons of FMEA and FTA approaches.

(11) Write an essay on MIL-HDBK-217.

4.12 REFERENCES

1. Smith, S.A., Service Reliability Measured by Probabilities of Outage, *Electrical World*, Vol. 103, 1934, pp. 371–374.
2. Dhillon, B.S., *Power System Reliability, Safety and Management*, Ann Arbor Science Publishers, Ann Arbor, Michigan, 1983.
3. Kececioglu, D., *Reliability Engineering Handbook, Vol. 1*, Prentice-Hall, Inc., Englewood Cliffs, New Jersey, 1991.
4. Shooman, M.L., *Probabilistic Reliability: An Engineering Approach*, McGraw-Hill Book Company, New York, 1968.
5. *Reliability, Maintainability and Supportability Guidebook*, SAE G-11 RMS Committee Report, Published by Society of Automotive Engineers (SAE), Inc., 400 Commonwealth Drive, Warrendale, Pennsylvania, 1990.
6. Henney, K., editor, *Reliability Factors for Ground Electronic Equipment*, McGraw-Hill Book Company, New York, 1956.
7. Dhillon, B.S., *Reliability and Quality Control: Bibliography on General and Specialized Areas*, Beta Publishers, Inc., Gloucester, Ontario, Canada, 1992.
8. Dhillon, B.S., *Reliability Engineering Application: Bibliography on Important Application areas*, Beta Publishers, Inc., Gloucester, Ontario, Canada, 1992.
9. Niebel, B.W., *Engineering Maintenance Management*, Marcel Dekker, Inc., New York, 1994.
10. Dhillon, B.S., A Hazard Rate Model, *IEEE Transactions on Reliability*, Vol. 28, 1979, p. 150.
11. Shooman, M.L., *Probabilistic Reliability: An Engineering Approach*, McGraw-Hill Book Company, New York, 1968.
12. Dhillon, B.S., Singh, C., *Engineering Reliability: New Techniques and Applications*, John Wiley and Sons, New York, 1981.
13. Dhillon, B.S., *Reliability Engineering in Systems Design and Operation*, Van Nostrand Reinhold Company, New York, 1983.
14. Elsayed, E.A., *Reliability Engineering*, Addison Wesley Longman, Inc., Reading, Massachusetts, 1996.
15. Ramakumar, R., *Engineering Reliability: Fundamentals and Applications*, Prentice-Hall Inc., Englewood Cliffs, New Jersey, 1993.
16. Sundararajan, C., *Reliability Engineering*, Van Nostrand Reinhold, New York, 1991.
17. MIL-HDBK-217, *Reliability Prediction of Electronic Equipment*, Department of Defense, Washington, D.C., 1978; available from the National Technical Information Service (NTIS), Springfield, Virginia 22161.

18. Countinho, J.S., Failure Effect Analysis, *Transactions of the New York Academy of Sciences*, Vol. 26, 1964, pp. 564–584.
19. Dhillon, B.S., Failure Modes and Effects Analysis—Bibliography, *Microelectronics and Reliability*, Vol. 32, 1992, pp. 719–731.
20. Vesley, W.E., Goldberg, F.F., Roberts, N.H., Haasl, D.F., *Fault Tree Handbank*, Report No. NUREG-0492, U.S. Nuclear Regulatory Commission, Washington, D.C., 1981.
21. Schroder, R.J., Fault Tree for Reliability Analysis, *Proceedings of the Annual Symposium on Reliability*, Institution of Electrical and Electronic Engineers (IEEE), New York, 1970.
22. MIL-HDBK-217, *Reliability Prediction of Electronic Equipment*, Department of Defense, Washington, D.C.; available from the National Technical Information Service (NTIS), Springfield, Virginia, 1978.
23. Dhillon, B.S., Viswanath, H.C., Bibliography of Literature on Failure Data, *Microelectronics and Reliability*, Vol. 30, 1990, pp. 723–750.

CHAPTER 5

Maintainability Engineering

5.1 INTRODUCTION

MAINTAINABILITY engineering is an engineering discipline that formulates a satisfactory combination of design features, maintenance facilities, and repair policies to acquire a given level of maintainability, as an operational requirement, at minimum life cycle costs. In contrast, maintainability is a design characteristic that, when effectively achieved, contributes to fast, efficient, and straightforward maintenance at the minimum life cycle cost. Since the term *maintenance* is often incorrectly interchanged with maintainability, the clarification of their true meanings is absolutely essential. Maintainability is design-related and, thus, considered during the design of a product; in contrast, maintenance is operation-related and refers to those activities performed after the product is functioning in the field to keep it in operational condition or to repair it to an operational state after experiencing a failure.

The roots of the maintainability discipline can be traced back to the beginning of this century; for example, in 1901 the Army Signal Corps contract document concerning the development of the Wright Brothers' well-known machine contained a clause that the airplane be "simple to operate and maintain" [1]. However, in its modern context, the beginning of maintainability goes back to the early 1950s as the result of various startling findings: (1) the U.S. Army's Eighth Air Force based in England during World War II discovered that only 30% of the heavy bombers stationed at one field were in operational readiness and the situation at other fields was similar [2]; (2) a study conducted by the U.S. Navy revealed that during maneuvers the electronic equipment was operative only 30% of the time; and (3) an army study reported that between two-thirds and three-fourths of the equipment was either out of commission or under repair [3]. In 1957, the Advisory Group on Reliability of Electronic Equipment

(AGREE) published the recommendations of its nine task groups. Mostly, the current standards on maintainability have their roots in those recommendations [4].

During the period from the late 1950s to the early 1960s, the U.S. Air Force launched a program to develop a systematic approach to maintainability. As the result, this effort led to the publication of the Air Force Maintainability Regulation 66-29 and specification MIL-M-26512. Subsequently, many related Department of Defense documents were released, including MIL-STD-778 (definition of maintainability terms) [5], MIL-STD-470 (maintainability program requirements) [6], MIL-STD-471 (maintainability demonstration) [7], MIL-HDBK-472 (maintainability prediction) [8], and AMCP 706-133 (maintainability engineering theory and practice) [1].

In 1984, the U.S. Air Force launched the Reliability and Maintainability (R&M) 2000 initiative with the purpose of making system/equipment R&M characteristics equal to performance, cost, and schedule in the development and decision process. Similarly, the other two services (i.e., army and navy) established their own R&M initiatives; in particular, during the 1980s the Army set a goal of halving the operational and support costs driven by reliability, maintainability, durability, and availability by the year 1991 [4].

This chapter discusses various different aspects of maintainability in engineering design.

5.2 MAINTAINABILITY DEFINITION, OBJECTIVES, AND PROGRAM ELEMENTS

Maintainability is defined as the probability that a failed item will be restored to its satisfactory operational condition within a specified time period when the maintenance is performed according to stated procedures and resources.

Maintainability engineering is performed to fulfill various objectives: to influence design to achieve ease of maintenance, to estimate the downtime for maintenance, to estimate system availability, to estimate the time and other resources needed for performing maintenance, and so on [9]. More sspecifically, the general principles of maintainability include eliminating or minimizing the need for maintenance; minimizing the frequency and complexity of required maintenance tasks; minimizing the levels of skill required to perform maintenance; providing the most suitable accessibility to items requiring frequent maintenance, replacement, removal, inspection, or adjustment; providing the means for efficient identification of item malfunctions; minimizing the necessity of special tools; maximizing interchangeability of items; providing for the efficient facilitation of transport such as towing, hoisting, lifting, and jacking; combining components of a system/equipment or function of a system/equipment into a removable assembly; providing for self-adjustment of parts subject to wear; and so forth [10].

Some of the elements of a maintainability program are to develop a maintainability program plan, conduct maintainability analysis, develop maintainability design criteria, perform design trade-offs, predict maintainability values, participate in design reviews, incorporate maintainability requirements in subcontractor specifications, demonstrate maintainability requirement achievements, and develop maintainability status reports [9].

5.3 MAINTAINABILITY IN DESIGN PHASE AND SUPPORT PROCESS

Usually, around 85–90% of owning and operating a system/equipment costs are determined by decisions taken prior to completing the detailed system/equipment design. Thus, the design activity of early development phases is extremely important in determining the product maintainability. It means that maintainability design evaluation is to be practiced right from the initial design phase to ensure that the designer carefully considers the performance of required maintenance on the product being designed. Since the design process can be basically divided into four phases—i.e., the concept development layout, detail design, assembly design, and installation—the maintainability considerations are discussed in each of these phases separately [4].

(1) Concept development layout phase: During this phase the basic configuration of the product is established, and the maintainability specialist/designer ensures the existence of sufficient access for all necessary maintenance activities, including appropriate amount of space for removal of parts, maintenance workers, and any tools or support equipment necessary. This is usually accomplished by evaluating layout design drawings and computer models developed by the design personnel. Also, maintainability specialists pay attention to items such as auxiliary system layout, partitioning of units, maintenance environment, and the layout susceptibility to maintenance damage.

(2) Detail design phase: During this phase emphasis is on the design of detail components that will ultimately form the product. The design work in this phase basically determines the maintenance details that will be carried out to support the higher level assemblies up to and inclusive of the end item. More specifically, a strong emphasis on maintainability is placed to avoid release of a design that needs very time-consuming overhauls, leading to an increase in maintenance time, a longer repair pipeline for parts, and

additional spares to achieve the same level of system/equipment/product availability. A careful consideration is also given to the support equipment during this phase, along with accomplishing maintainability analysis, predictions, allocations, and models.

(3) Assembly design phase: This phase is concerned with combining components into assembly or integrated into systems. Maintenance needs requiring some special knowledge and training at the intermediate level are determined along with the assembly/disassembly (for maintenance) of larger, more complex pieces of the product.

(4) Installation phase: In this phase all the assemblies/systems are put together to create the final product. Also, the details of the maintenance tasks that usually can be acted upon by individuals with a minimal level of training are determined. It is to be noted that, even though such tasks were determined by earlier design phases, there remains a possibility of being negated by unforeseen circumstances. Nevertheless, the major emphasis of the maintainability activity is to ensure that the features incorporated during the previous phases of design are not nullified.

Usually, the support process is established prior to putting a system/equipment into service and it matures over time according to demands. Because the great majority of the maintainability problems or successes cannot be ascertained without having sufficient user data, over the years many extensive data systems have been developed for keeping track of the system/product maintainability. Such data systems are useful in revealing the realized maintainability through the measures of manpower, time, demand for spares, and so on. An effective data collection system can provide invaluable information to maintainability specialists and others; thus, such a system must provide information on many different areas: What failed? How and why did it fail? How was it repaired? Who repaired it? and How can future failures be prevented? All in all, it should be remembered that, like many other engineering disciplines, maintainability is also a continuing process.

5.4 MAINTAINABILITY MEASURES

Because maintainability is a system/equipment/product design characteristic, it addresses issues such as the ease, accuracy, timeliness and cost-effectiveness of maintenance actions. Maintainability may be measured in terms of a combination of maintenance cost, labor rates, frequency of maintenance, labor hours, elapsed times, and so on. The quantitative assessment of system/equipment/product maintainability is facilitated by such measures. Obviously, the main

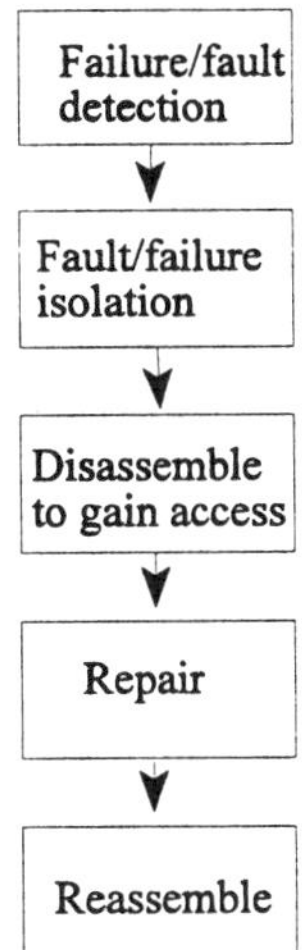

Figure 5.1 Steps for a typical corrective maintenance or repair cycle

objective of calculating maintainability measures is to influence design and ultimately produce a cost-effectively maintainable product. Some of the maintainability measures are presented below separately.

5.4.1 MEAN TIME TO REPAIR (MTTR)

This probably is the most widely used maintainability measure, and it measures the elapsed time needed to conduct a maintenance activity. Subsequently, MTTR is used to compute system/equipment/product availability and downtime. It must be remembered that MTTR is also referred to as mean corrective maintenance time. A typical corrective maintenance or repair cycle is composed of steps such as shown in Figure 5.1. MTTR is expressed as

$$\text{MTTR} = \left(\sum_{j=1}^{k} \lambda_j T_j \right) \Bigg/ \sum_{j=1}^{k} \lambda_j \tag{5.1}$$

where

k = the total number of items
T_j = the corrective maintence or repair time of the jth item
λ_j = the constant failure rate of the jth item

Even though MTTR may be calculated using Equation (5.1), normally the corrective maintenance times are described by the following three probability distributions [11]:

- **exponential:** This sometimes is assumed for electronic equipment having an effective built-in test capability along with a rapid remove and replace maintenance concept.
- **normal:** This is mainly assumed for mechanical or electromechanical equipment usually having a remove and replace maintenance concept.
- **log-normal:** This is often assumed for electronic equipment without a built-in test capability. Also, it can find applications to electromechanical equipment having widely variant individual repair times.

5.4.2 MEAN PREVENTIVE MAINTENANCE TIME

Preventive maintenance is performed to keep a system/equipment/product at a specified performance level and includes functions such as inspections, calibration, and tuning. The mean preventive maintenance time is expressed by [11]

$$T_{mp} = \left(\sum_{i=1}^{k} T_{pi} f_i\right) \Bigg/ \sum_{i=1}^{k} f_i \qquad (5.2)$$

where

T_{mp} = the mean preventive maintenance time
k = the total number of preventive maintenance tasks
T_{pi} = the time taken to perform the ith preventive maintenance task
f_i = the frequency of the ith preventive maintenance task in actions per system/equipment/product operating hour

It is to be noted that T_{mp} does not include administrative and logistics delay times but only the downtime expended for performing preventive maintenance.

5.4.3 MAXIMUM CORRECTIVE MAINTENANCE TIME

This is the maximum time to repair for a specified percentile measure, the time required to carry out the given percentage of all identified potential repairs. The frequently used percentiles are the 90th and the 95th. More specifically, this measure specifies the upper limit on the percentage of tasks allowable to bypass a given repair time period. For example, if the specified maximum corrective maintenance time is 3 hours at the 90th percentile, then no more than 10% of the repair activity times are expected to be above 3 hours. Nevertheless, the computation depends on the probability density assumed to describe the corrective maintenance time. Thus, for normal and log-normal distributions, the maximum corrective maintenance time is given by Equations (5.3) and (5.4), respectively [9, 11].

$$MT_n = T + Z\sigma \tag{5.3}$$

where

MT_n = the maximum corrective maintenance time for the normal distribution
T = the average of individual times to repair
σ = the standard deviation of distribution of times to repair
Z = the percentile factor

$$MT_{n\ell} = \text{antilog}\,[\log t + Z\sigma_{n\ell}] \tag{5.4}$$

where

$$\log t = \left(\sum_{i=1}^{h} \log t_i\right) k \tag{5.5}$$

In turn, t_i is the ith corrective maintenance time and k the total number of t_i.

$$\sigma_{n\ell} = \sqrt{\left[\sum_{i=1}^{k} (\log t_i)^2 - \left(\sum_{i=1}^{k} \log t_i\right)^2 \Big/ k\right] \Big/ (k-1)} \tag{5.6}$$

The values of Z for the 90th and 95th percentiles are 1.28 and 1.65, respectively.

5.4.4 PROBABILITY OF REPAIR WITHIN ALLOWABLE DOWNTIME

This is concerned with carrying out a repair within a specified time interval $t(s)$. Thus, the probability is expressed by [9]

$$P(S) = 1 - e^{-[t(s)/\text{MTTR}]} \tag{5.7}$$

where

MTTR = the mean time to repair or the expected downtime
$t(s)$ = the allowable downtime

The number of failures, NR, that cannot be repaired within the allowed downtime is expressed by

$$NR = (\lambda T)e^{-[t(s)/\text{MTTR}]} \tag{5.8}$$

where

λ = the constant failure rate expressed in failures per hour
λT = the mean number of failures occurring during a mission time T

Similarly, the number of failures, N, that can be repaired within the allowed downtime is given by

$$N = (\lambda T)P(S) \tag{5.9}$$

5.5 MAINTAINABILITY FUNCTIONS

When a repair/maintenance action starts at time $x = 0$, it is normally beneficial to predict the probability of completing the repair in a time x. For this very purpose, the maintainability functions are used. The maintainability function is defined by

$$MF(x) = \int_0^x f(x)\mathrm{d}x \tag{5.10}$$

where

$MF(x)$ = the cumulative distribution function for a given repair/maintenance time distribution; also denotes the probability of accomplishing a repair within a specified time interval $[0, x]$
x = the allowable downtime for a product/item/equipment
$f(x)$ = the product/item/equipment repair time probability density functions

Maintainability functions for several different repair/maintenance time probility density functions are presented below [1].

Exponential

$$MF(x) = 1 - e^{-\mu x} \tag{5.11}$$

where μ is the constant repair/maintenance rate of the product/equipment/item.

Weibull

$$MF(x) = 1 - e^{[-(x/k)^b]} \tag{5.12}$$

where

k = the scale parameter
b = the shape parameter

Gamma

$$MF(x) = \frac{\mu^m}{\Gamma(m)} \int_0^x t^{m-1} e^{-\mu t} \mathrm{d}t \tag{5.13}$$

where

m = the shape parameter
μ = the scale parameter
$\Gamma(m)$ = the gamma function and is defined as

$$\Gamma(m) = \int_0^\infty t^{m-1} e^{-t} \mathrm{d}t \tag{5.14}$$

Some of the special case values of Equation (5.14) are as follows:

- $m = 1,$ $\Gamma(1) = 1$
- $m = 0.5,$ $\Gamma(0.5) = \sqrt{\pi}$
- $m = 0, 1, 2, \ldots$ $\Gamma(m + 1) = m!$

Normal

$$MF(x) = \frac{1}{\sigma\sqrt{2\pi}} \int_{t=-\infty}^{x} \exp\left[-\frac{1}{2}\left(\frac{x-\theta}{\sigma}\right)^2\right] \mathrm{d}x \tag{5.15}$$

where θ is the mean repair/maintenance time and is defined for n data points as

$$\theta = \sum_{i=1}^{n} x_i / n \tag{5.16}$$

where x_i is the ith maintenance/repair time. Similarly, σ is the standard deviation and is expressed as

$$\sigma = \sqrt{\sum_{i=1}^{n} (x_i - \theta)^2 / (n-1)} \tag{5.17}$$

Log-normal

$$MF(x) = \frac{1}{\sigma\sqrt{2\pi}} \int_0^\infty \exp\left[-\frac{1}{2}\left(\frac{\ln x - \beta}{\sigma}\right)^2\right] \mathrm{d}x \tag{5.18}$$

where β is the average of the natural logarithms of n repair/maintenance times and is expressed by

$$\beta = \left(\sum_{i=1}^{n} \ln x_i \right) \Big/ n \tag{5.19}$$

where

x_i = the ith maintenance/repair time
σ = the standard deviation with which the natural logarithm of the repair/maintenance times are spread around the mean β and is defined as

$$\sigma = \sqrt{\sum_{i=1}^{n} \left(\ln x_i - \beta \right)^2 / (n-1)} \tag{5.20}$$

5.6 SPARE PART QUANTITY AND MAINTENANCE LABOR-HOURS PER OPERATING HOUR PREDICTIONS

The determination of spare part quantity is a function of various factors: the number of items used in the product/system/equipment, the reliability of the items under consideration, the probability of having a spare available when required, and so forth. The spare part quantity may be determined by using the following formula [11]:

$$PS = \sum_{i=0}^{i=N} R[(-1) \ln R]^i / i! \tag{5.21}$$

where

N = the number of spare parts/items carried in stock
PS = the probability of having a spare of a certain item available when needed
R = the part/parts reliability and is expressed as

$$R(t) = e^{-M\lambda t} \tag{5.22}$$

where

$R(t)$ = the part/parts reliability at time t
λ = the constant failure rate of a part
M = the number of parts used of a specific type

Since *PS* directly impacts the quantities of spare parts to be procured, it may be called a safety factor. The higher value of the safety factor is achieved by the greater number of spares that lead to a higher cost of item acquisition and inventory maintenance.

Maintenance labor-hours per operating hour, $T_{m\ell}$, is one of the most commonly used maintainability parameters and measures labor-hours required to perform a maintenance task. In addition, this parameter is used to estimate maintenance costs and for personnel planning [9]. $T_{m\ell}$ is defined by

$$T_{m\ell} = t_C(A + B)/T_{OC} \tag{5.23}$$

$$A = \sum_{i=1}^{k}(\lambda_i \mathrm{MTTR}_i MC_i) \tag{5.24}$$

$$B = \sum_{j=1}^{k}(NP_j \mathrm{MTPM}_j MC_j) \tag{5.25}$$

where

t_C = the calendar time duration under consideration
T_{OC} = the system/equipment operating time during calendar time t_c
k = the total number of elements under consideration
MTTR_i = the ith element mean time to repair expressed in hours
λ_i = the ith element constant failure rate
MC_i = the mean number of maintenance personnel needed to repair the ith element
NP_j = the total number of times annually preventive maintenance is needed for the jth element
MTPM_j = the mean time needed to carry out preventive maintenace on the jth element, expressed in hours
MC_j = the mean number of maintenance personnel needed to carry out maintenance on the jth element

5.7 RELIABILITY-CENTERED MAINTENANCE

Reliability-centered maintenance (RCM) may be described as a systematic logic or methodology used to highlight maintenance tasks to realize the inherent equipment/system/product reliability by expending minimum resources [12]. More specifically, RCM is a disciplined approach for developing a focused and cost-effective preventive maintenance program and control plan for a system/equipment/product/process. It is usually best to initiate RCM in the early system/equipment/product design process, and the technique evolves as the design, development, and deployment activities progress [11]. In addition, RCM can also be applied to evaluate preventive maintenance programs for already existing systems.

The beginning of the RCM methodology goes back to the 1960s due to the efforts of the commercial airline operators and manufacturers through the U.S. Air Transport Association (ATA) [12]. In those years, a maintenance steering group (MSG) was established and that group ultimately produced a document entitled "747 Maintenance Steering Group Handbook: Maintenance Evaluation and Program Development," MSG-1 [13]. Because this document was focused on one particular type of aircraft (i.e., Boeing 747), a generalized version entitled "Airline/Manufacturer Maintenance Program Planning Document," MSG-2 [14] of it was published. Thus, this document became applicable to other wide-body aircrafts known as DC-10 and L-1011. In fact, MSG-2 served as a practical guide to RCM practice for over a decade. In 1978, the MSG-2 was refined and published by the United Airlines as Report No. A066-579 entitled "Reliability Centered Maintenance" [11].

In 1980, ATA published MSG-3 [15], again entitled "Airline/Manufacturers Maintenance Planning Document." This document covers maintenance programs for aircraft such as Boeing 756, Boeing 767, Concorde, and A-300 [12]. In 1993, MSG-3 was renamed "Airline/Manufacturer Maintenance Program Development Document" [11, 16].

Some of the important current documents concerning RCM are as follows:

- AMC-P-750-2: Guide to Reliability-Centered Maintenance [17]
- MIL-STD-2173 (AS): Reliability-Centered Maintenance Requirements for Naval Aircraft, Weapon Systems, and Support Equipment [18]
- Book: Reliability-Centered Maintenance by J. Moubray [19]

Reference [17] specifically states RCM objectives as follows:

- Develop priorities in design that facilitate preventive maintenance.
- Plan preventive maintenance tasks for the purpose of restoring reliability and safety to their inherent levels in the event of the occurrence of system/equipment deterioration.
- Collect appropriate data for improving the design of items with unsatisfactory inherent reliability.
- Achieve the above goals at a least total cost, including residual failure costs and maintenance costs.

The RCM process includes six essential activities: (1) establishing required part criticality input data, (2) making use of the part criticality data to the

RCM decision logic process for choosing appropriate maintenance tasks, (3) documenting the final decisions, (4) determining intervals associated with tasks, (5) implementing the final RCM task decisions and intervals, (6) reviewing results and refining as necessary.

5.7.1 RCM APPLICATION FOR DEVELOPMENTAL SYSTEMS/EQUIPMENT/PRODUCTS

This section discusses RCM tasks during the product/system/equipment life cycle [12]:

- concept exploration: In this phase the RCM tasks and their interfacing with the integrated logistic support process are planned. Also, the maintenance concept is developed, including the highlighting of the capability required to maintain reliability and safety parameters entrenched during design.
- full-scale development: There are various tasks performed during this phase, including updating of RCM documentation at the early stage of this phase, performance of complete RCM logic analysis, performance of maintenance analysis for determining maintenance tasks required for each repairable item, evaluation of the overall maintenance plan adequacy with respect to appropriate test plans, and monitoring of RCM implementation by the review team.
- production and deployment: Some of the tasks performed during this phase are developing maintenance procedures for maintenance-significant items according to the RCM logic analysis, incorporating maintenance requirements derived from RCM analyses into technical publications, determining appropriate intervals for each maintenance task, instituting a sustaining engineering program, demonstrating fully the maintenance concept/plan, and assessing feedback data according to desirability.

5.8 BACKGROUND DATA FOR MAINTAINABILITY DESIGN AND MAINTAINABILITY DESIGN CHARACTERISTICS

There are various kinds of maintainability input data needs during system/equipment/product design. Often, it is experienced that the maintainability-related input requirements are incomplete and in primitive form. To use this information during the design phase, the concerned maintainability specialist should seek answers to the following [1]:

- the purpose of designing the system/product/equipment and its operational requirements
- maintenance objectives
- environmental and policy considerations (e.g., maintenance policies, operational and resource constraints, applicable maintainability documents, and integrated logistic support concept)
- mission profiles, preventive versus corrective maintenance, and logistic endurance factors
- the way the product/system/equipment is to be supported: depth of maintenance, repair/discard/replace policies, periodic test and checkout, and the use of standby redundancy
- supporting manpower: operators and organization repairmen

There are many basic system/equipment/product characteristics that should be emphasized during design for maintainability. These include test points, displays, standardization, interchangeability, modular design, safety, human factors, skills, illumination, controls, test equipment, labeling and coding, tools, accessibility, installation, lubrication, weight, adjustments and calibrations, manuals, work environment, training needs, size and shape, ease of removal/replacement, circuit breakers, covers and doors, test hookups and adapters, connectors, failure indication (location), servicing equipment, and operability. The most commonly mentioned maintainability characteristics by the people involved with maintainability are controls, tools, displays, labeling and coding, accessibility, test points, test equipment, safety, connectors, handles and handling, mounting and fasteners, cases, covers, doors, manuals, checklists, charts, and aids.

It is to be noted that system/equipment maintainability design characteristics include those features and design factors that will help to decrease equipment/system downtime. In addition, important factors to consider are as follows:

- decrease in support costs
- ease of performing maintenance
- reduction in the number of preventive and corrective maintenance tasks to be performed to a minimum level
- minimization of the logistical burden through decrease in need for maintenance and support resources (e.g., levels of skills, number of people, repair parts, special maintenance facilities, and support equipment)

Selective specific features and their effect on maintainability design are discussed below separately [1].

5.8.1 PACKAGING

Packaging is a very important factor in equipment/product/system maintainability. In fact, factors such as layout of parts, components, and assemblies, their mounting, access, and ease of removal or repair contribute significantly to ease of maintenance and maintenance downtime. Fortunately, most of these items can be located and packaged in various ways with varying degrees of success. Some of the factors that should be considered in packaging related decisions are listed below:

- standardization needs
- modularization requirements
- operating stress
- accessibility requirements
- vibration
- reliability factors
- accessibility preferences
- temperature
- manufacturing requirements
- built-in test requirements
- test point related requirements
- safety
- characteristics peculiar to each item, including weight, size, fragility, servicing, heat dissipation, and clearance

5.8.2 STANDARDIZATION

Standardization may be described as a design feature for restricting to the least amount possible the variety of parts/items that will satisfy the majority of the requirements associated with a system/product/equipment. The standardization leads to reduction in both the procurement and the support costs associated with a system/equipment/product in addition to increasing in its reliability and maintainability. In order to achieve full benefits of standardization, it must be practiced at all stages of design, as well as to items within the supply system. Some of the primary goals of standardization are as follows:

- Minimize the use of different types of parts/assemblies/items.
- Minimize the use of different models and makes of equipment.
- Maximize the use of interchangeable parts/components/items.

- Increase the use of common parts/components/items in different equipment/systems/products.
- Simplify and reduce part coding, numbering practices, and so on.
- Use standard off-the-shelf items/parts/components as much as possible.

Under the circumstances of performing standardization to the maximum in system/equipment/product design, there are specific important advantages gained by the support activities needed for the finished system/equipment/product. Some of those advantages include reduction in the varieties and quantities of support test equipment, reduction in the types and the quantities of spares of increased reliability, reduction in training requirements for support personnel, reduction in the number and types of support personnel required, and reduction in the need for support facilities.

5.8.3 INTERCHANGEABILITY

Interchangeability, a maintainability design factor, is quite closely related to standardization and is realized through standardization. It may simply be described as a design policy whereby any specified part/item can be substituted in an assembly/equipment/system for any similar part/component/unit in accordance with the standardization principles. For the effectiveness of interchangeability, liberal tolerances are quite essential. In any case, to achieve maximum interchangeability of parts/components/units, the equipment designers must insure the factors such as follows [1]:

- existence of functional interchangeability in situations wherever physical interchangeability is identified as a design characteristic
- avoiding differences in size, shape, and mounting
- nonexistence of physical interchangeability in situations whenever functional interchangeability is not desired
- design embracing the principle for functional interchangeability wherever total (i.e., functional and physical) interchangeability is impracticable; in addition, existence of adapters to make possible physical interchangeability as much as practical
- modifications of parts/components/units not changing the mounting and connecting mechanisms
- existence of total interchangeability for all parts/components/units intended to be identical
- existence of sufficient information on in-job instructions and on identification plates to allow users to decide with confidence whether or not similar items/parts/components are interchangeable

Basically, the benefits gained from effective interchangeability are virtually the same as that of standardization.

5.8.4 CHECKLISTS

Checklists are an important aid for designers to insure that all essential design factors influencing maintainability characteristics associated with a system/equipment/product receive adequate consideration. Thus, checklists help in both design and review situations. Checklists may be used by the following professionals for three different purposes:

(1) Maintainability engineers: They can use them to predict maintainability.
(2) System engineers/project managers: They can use them to review design during different product/system/equipment life cycle phases.
(3) Design and maintainability engineers: They can use them to consider the influence of certain maintainability design features.

5.9 GENERAL DESIGN GUIDELINES FOR MAINTAINABILITY AND COMMON MAINTAINABILITY-RELATED DESIGN ERRORS

There are many important tasks that have to be performed by professionals involved with maintainability during design. However, Reference [20] presents a number of important general design guidelines concerning maintainability: design for minimum tools, adjustment, and maintenance skills; design for safety; provide for visual inspection; make use of standard interchangeable parts as much as possible; provide troubleshooting techniques, test points, and handles on heavy parts/components/units; use color-coding, captive-type chassis fasteners and plug-in, rather than solder-in, modules; label units and avoid the use of large cable connectors and group subsystems (e.g., power supply components) so that they can be located and identified easily.

In many instances, engineering designers have repeatedly made design errors that affect product/equipment/system maintainability. Some of those common design errors were as follows [20]:

- placed adjustments out of reach of a person's arm
- overlooked to provide sufficient room for a person's gloved hand to make necessary adjustments
- failure to install handles or appropriate handles
- placed removable parts/items to make it impossible to remove them without taking the entire unit from its case or without first removing other parts
- access doors installed with numerous small screws

- placed adjusting screws in locations difficult to find by the repair persons
- installed low-reliability parts beneath other parts, thus making maintenance people disassemble many components to reach them
- located adjusting screws close to a hot component or to an exposed power-supply terminal
- located screwdriver adjustments beneath modules, thus making it difficult for repair persons to reach them
- used chassis and cover plates that drop, in the event of removing the last screw
- used low-reliability test equipment that caused false reports of product failure
- placed fragile parts/components just within the bottom edge of the chassis, where repair persons may put their hands
- made sockets and connectors the same for modules, consequently leading to the installation of wrong modules
- screwed together subassemblies in such a manner so that repair people fail to differentiate what is being held by the screw

5.10 PROBLEMS

(1) Define the following three terms:
- preventive maintenance
- maintainability
- maintainability engineering

(2) Discuss the historical developments in the discipline of maintainability engineering.

(3) Discuss at least three measures of maintainability.

(4) Write down the probability density function for the Rayleigh distribution and then develop an expression for the maintainability function.

(5) What is the maintainability function? Discuss its relationship to the reliability function.

(6) Define the term *reliability-centered maintenance*. What are the principal objectives of the reliability-centered maintenance?

(7) List at least 10 important product characteristics that should be carefully considered during design for maintainability.

(8) Describe in detail the following two subjects:
- standardization
- interchangeability

(9) Discuss at least 10 common design errors that affect maintainability.

5.11 REFERENCES

1. *Engineering Design Handbook: Maintainability Engineering Theory and Practice*, AMCP 706-133, 1976, prepared by the U.S. Army Material Command, Department of the Army, 5001 Eisenhower Ave., Alexandria, VA.
2. Moss, M.A., *Minimal Maintenance Expense*, Marcel Dekker, Inc., New York, 1985.
3. Shooman, M.L., *Probabilistic Reliability: An Engineering Approach*, McGraw-Hill Book Company, New York, 1968.
4. *Reliability, Maintainability and Supportability Guidebook*, SAE G-11, 1990, published by Society of Automotive Engineers (SAE), Inc., 400 Commonwealth Drive, Warrendale, PA.
5. MIL-STD-778, *Definition of Maintainability Terms*, Department of Defense, Washington, D.C.
6. MIL-STD-470, *Maintainability Program Requirements*, Department of Defense, Washington, D.C., March 1966.
7. MIL-STD-471, *Maintainability Demonstration*, Department of Defense, Washington, D.C., February 1966.
8. MIL-STD-472, *Maintainability Prediction*, Department of Defense, Washington, D.C., May 1966.
9. Kraus, J.W., Maintainability and Reliability, in *Handbook of Reliability Engineering and Management*, McGraw-Hill Book Company, New York, 1988, pp. 15.1–15.38.
10. Niebel, B.W., *Engineering Maintenance Management*, Marcel Dekker Inc., New York, 1994.
11. Blanchard, B.S., Verma, D., Peterson, E.L., *Maintainability*, John Wiley & Sons, Inc., New York, 1995.
12. Anderson, R.T., Neri, L., *Reliability-Centered Maintenance*, Elsevier Applied Science, London, 1990.
13. "747 Maintenance Steering Group Handbook: Maintenance Evaluation and Program Development," MSG-1, Air Transport Association, Washington, D.C., 1968.
14. "Airline/Manufacturer Maintenance Program Planning Document," MSG-2, Air Transport Association, Washington, D.C., 1970.
15. "Airline/Manufacturer Maintenance Program Planning Document," MSG-3, Air Transport Association, Washington, D.C., 1980.
16. "Airline/Manufacturer Maintenance Program Development Document," MSG-3, Air Transport Association, Washington, D.C., 1993.
17. "Guide to Reliability-Centered Maintenance," AMC-P-750-2, Department of Defense, Washington, D.C., 1985.
18. "Reliability-Centered Maintenance Requirements for Naval Aircraft, Weapon Systems, and Support Equipment," Department of Defense, Washington, D.C.
19. Moubray, J., *Reliability-Centered Maintenance*, Butterworth-Heinemann, London, 1991.
20. Pecht, M., editor, *Product Reliability, Maintainability, Supportability Handbook*, CRC Press, Boca Raton, Florida, 1995, pp. 191–192.

CHAPTER 6

Safety Engineering

6.1 INTRODUCTION

SAFETY is becoming a very crucial component in engineering design as product liability suits are on the increase. There were several different aspects of product liability actions according to a survey of court actions in 1973: strict liability accounted for 42% of the cases reviewed, breach of warranty accounted for 40% of the cases, and negligence was the reason for 18% of the cases.

In 1980, over 10 million Americans suffered a disabling injury, in dollars and cents translating into $83.2 billion, i.e., roughly 8% of the gross national product of the United States.

The concern for safety has existed for many centuries: Pliny the Elder (AD 23-79) [1] pointed to fumes from lead and the dust from mercury ore grinding; Hippocrates identified lead poisoning in the fourth century BC [2]. In the sixteenth century, George Bauer (1492–1555), also known as Agricola, "the father of mineralogy," wrote a 12-volume series on mining and metallurgy, which described mining and its associated hazards in detail.

Bernadino Ramazzini (1633–1714), who should probably be called the "father of occupational health and safety," systematically described occupational diseases. In 1775, Percival Pott identified the link between soot and cancer. The invention of the printing press by Gutenberg in Germany around 1450 may be called the beginning of the industrial revolution–related safety problems. In response to the request of the Society for the Prevention of Accidents in Coal Mines, Humphrey Davy developed the minor's safety lamp.

Other major milestones in the development of the safety field are as follows [3, 4]:

- 1736: Benjamin Franklin established the first volunteer fire department, known as the Union Fire Company of Philadelphia.
- 1866: The National Board of Fire Underwriters was formed.
- 1868: The first barrier safeguard was patented.
- 1869: The Pennsylvania legislature passed a mine safety law requiring two exits from all mines. Also, George Westinghouse developed a brake based on compressed air that helped to improve safety quite significantly in railroad travel.
- 1878: Edward Atkinson developed less hazardous production equipment and techniques.
- 1892: The first safety program in American industry was established.
- 1912: The Association of Iron and Steel Electrical Engineers (AISEE) organized the first meeting of the Cooperative Safety Congress in Milwaukee.
- 1913: National Council of Industrial Safety was formed and changed its name to National Safety Council in 1915.
- 1931: H. W. Heinrich [5] published a book on industrial safety entitled *Industrial Accident Prevention.*
- 1947: A paper entitled "Engineering for Safety," presented to the Institute of Aeronautical Sciences, emphasized the importance of designing safety into airplanes [6].

Since 1947, many other important events concerning safety have occurred, and a vast number of publications on the subject have also appeared [7]. This chapter discusses different aspects of safety, with emphasis on system/product safety.

6.2 SAFETY ENGINEERING GOALS AND SYSTEMS/PRODUCT SAFETY ENGINEERING

Although the overall goal of safety engineering is to reduce accidents and control or eliminate hazards in the workplace, realistically, it should concentrate on the following aims [3]:

- annual reductions in the occurrence of number of accidents
- annual reduction in workers' compensation insurance claims

The decades of the 1960s and 1970s witnessed rapid changes in industrial technologies and an increase in production process complexity, consequently forcing the development of better approaches and procedures for accident and injury prevention. This resulted in terms such as "systems safety engineering" and "systems safety analysis." It must be remembered that "systems safety analysis or engineering" was initially developed as a "product reliability" technique but at a later stage was used quite widely in the aerospace industrial sector.

Nowadays, manufacturers are being held legally liable for damages caused by their defective products in a vast number of cases. For example, in 1978 in the United States alone, over 1 million damage suits were filed [8]. The passage of the Consumer Products Safety Act on October 12, 1972, appeared to have been a contributory factor in a large number of damage suits. Another important factor for a greater attention on safety during product design, manufacture, testing, and distribution has been the establishment of the Consumer Products Safety Commission, which has the authority to ban the unsafe products from the marketplace. All in all, factors such as these have helped to increase the importance of having an effective product safety program.

6.3 HAZARD CLASSIFICATIONS AND COMMON MECHANICAL INJURIES

In order to produce safety effective products, it is of paramount importance to identify hazards during the early stages of design. Because there could be many different sources of hazards, the main classifications are as follows [9]:

- kinematic hazards
- environmental hazards
- electrical hazards
- human factors hazards
- energy hazards
- misuse- and abuse-related hazards

Locations where components or parts come together while moving and lead to possible cutting, pinching, or crushing of any object caught between them belong to the kinematic hazard classification.

The environmental hazards may be classified into two groups: internal and external. In the case of internal hazards, the changes in the surrounding environment lead to internally damaged product. For example, a drop in temperature below 32°F surrounding a water-cooled internal combustion engine will result in an internally damaged engine. Also, during the design phase to minimize or eliminate internal hazards, the product designer must carefully consider factors such as extremes of temperatures, vibrations, illumination level, atmospheric contaminants, electromagnetic radiation, and ambient noise levels.

The external hazards are the hazards posed by the proposed product during its life cycle, for example, service-life operation-related hazards, maintenance-related hazards, and disposal-related hazards.

The electrical hazards are a very important group of hazards. Two principal electrical hazards are the shock hazard and the electrocution hazard. Also, the major electrical hazard to property is from electrical faults, often referred to as short circuits.

The human factors hazards are associated with poor design with respect to the human element. The product or equipment design must carefully take into consideration appropriate human factors to avoid related hazards. These human factors include height, weight, intelligence, visual acuity, computational ability, physical strength, eduction, visual angle, and length of reach. Usually, for a given human factor, the common requirement is that a product or equipment be usable by a person between the 5th and the 95th percentile.

The energy hazards may be divided into two categories: potential energy and kinetic energy. The potential energy hazards are associated with components that store energy. Some examples of such components are springs, compressed-gas receivers, electric capacitors, and counterbalancing weights. The stored energy can cause injury when released suddenly; thus, this type of hazard is particularly important during the servicing of the equipment.

The kinetic energy hazards are associated with components that have energy because of their motion. Such components include fan blades, flywheels, and loom shuttles. Because these components have large amounts of kinetic energy, any object interfering with their motion could be damaged substantially.

Misuse- and abuse-related hazards are associated with product usage by humans. Sometime misuse of products can lead to serious injury, and product designers must carefully consider the misuse of products under design so that their misuse is either minimized or is eliminated altogether. Poor operating practices or lack of proper maintenance are normally the causes of product abuse and ultimately can lead to hazardous situations.

In an industrial setting, people interact with machines to perform tasks such as drilling, cutting, punching, shaping, stapling, chipping, and abrading. Some of the common injuries occurring as the result of performing such tasks are as follows [4]:

- cutting- and tearing-related injuries: These types of injuries occur when a body part comes in contact with a sharp edge. The severity of a cut or a tear depends upon the degree of damage to the skin, veins, arteries, muscles, and so forth.
- crushing-related injuries: These types of injuries occur when a body part gets caught between two hard surfaces moving progressively together and thus crushing anything between them. Alternatively, body parts can

also get crushed through other means: a hammer hitting a finger, a heavy object falling on a foot, or something similar.

- shearing-related injuries: These types of injuries are associated with shearing processes. In manufacturing, power-driven shears are widely used to perform tasks such as severing paper, plastic, elastomers, and metal. In the past, tragedies such as amputation of fingers/hands have occurred with the use of such machines.
- breaking-related injuries: These types of injuries are usually associated with machines used to deform engineering materials. A break in a bone is often referred to as a fracture, and fractures are classified into many categories: simple, compound, complete, incomplete, transverse, oblique, and comminuted.
- puncturing-related injuries: These types of injuries occur when an object penetrates straight into the body and pulls straight out, thus causing a wound to the body part. Normally, these types of injuries in the industrial sector are associated with punching machines because they have sharp tools.
- straining- and spraining-related injuries: In the industrial setup, there are numerous situations where the likelihood of straining of muscles or spraining of ligaments is possible. More specifically, a strain injury is the result of overstretched or torn muscles; in contrast, a sprain injury is caused by torn ligaments in a joint.

6.4 SAFETY IN SYSTEM LIFE CYCLE

Since the system/product may be divided into many phases, there are various safety-related actions specifically suitable for each of those phases. The five commonly accepted system/product life cycle phases are concept, definition, development, production, and deployment [6]. If the safety factor is considered carefully during each of those five phases, there is a strong chance that the end product will be user friendly with respect to safety. The appropriate safety measures for each of these five phases are discussed below.

6.4.1 CONCEPT PHASE

During this phase, past data and future technical projections are utilized to provide a basis for the system/product under consideration. In addition, concerning critical issues are examined, along with identification and evaluation of safety problems and their impacts. A useful analytical tool used to bring out the hazards during the concept phase is known as preliminary hazard analysis

(PHA). The main purpose of performing PHA is to have initial hazard analysis results in the development of the system/product and then take appropriate corrective measures. Nevertheless, at the close of this phase, questions such as those listed below should be asked:

- Are the design hazards identified and evaluated to develop hazard controls?
- Are the risk analyses initiated to develop the means of hazard control?
- Have the basic safety design needs for the phase developed so that the next (definition) phase can be started?

6.4.2 DEFINITION PHASE

This phase provides for verification of the initial design and engineering of the product/system under consideration. PHA is updated along with subsystem hazard analysis (SSHA) initiation and ultimate integration into the system hazard analysis (SHA). To examine specific known hazards and their associated effects, the fault tree analysis (FTA) and fault hazard analysis (FHA) may also be employed. One or more safety analysis techniques may be required to identify items such as specification of safety design requirements, safety equipment, and preliminary development of safety test plans and needs. All in all, system definition will initially lead to the acceptance of a desirable general design, even though at this stage not all hazards will be known because of the design incompleteness.

6.4.3 DEVELOPMENT PHASE

In this phase, integrated logistics support, producibility engineering, environmental impact, and operational use studies are added to the definition phase effort. The comprehensive operating hazard analysis (OHA) to examine man–machine hazards is conducted using prototype analysis and testing results. Also, since the design will be more complete, the PHA is developed further. The completion of hazard analysis, safety testing results, and complying with safety design criteria form a basis for making correct go/no-go decisions on the design under consideration.

6.4.4 PRODUCTION PHASE

During this phase, the developed design of the previous phase is produced in multiple units, and the monitoring at this stage by safety personnel is quite crucial. Depending upon the complexity of the product under production, it may be necessary to assign safety personnel to actual manufacturing to witness tests involving safety and to verify the performance of crucial tests leading to the ultimate product acceptance. The system/product safety engineering report (SSER) is prepared as the result of the production phase inputs, and the report itself highlights and documents the hazards associated with the end product/system.

6.4.5 DEPLOYMENT PHASE

During this phase, the product/system is being used in the field, and the data on failures, accidents, incidents, and so on are accumulated. As the result of the field use, if any changes to the product/system are made, the safety personnel carefully review such changes and update earlier safety analyses as considered appropriate.

6.5 SAFETY ANALYSIS METHODS

Over the years, many safety analysis methods have been developed, basically to aid decision makers and to identify safety-related weaknesses. Such methods include preliminary hazard analysis (PHA), hazard and operability review (HAZOP), technique of operations review (TOR), risk analysis, failure mode and effects analysis (FMEA), and fault tree analysis (FTA) [4]. Each of these methods is discussed below separately.

6.5.1 PRELIMINARY HAZARD ANALYSIS

PHA is basically a conceptual design phase approach, and therefore, it is a relatively unstructured method because of the unavailability of definitive functional flow diagrams, drawings, and so on. PHA is a useful tool to take early steps to identify and eliminate hazards, because of unavailability of all the desirable data. The results of PHA are quite useful in serving as a guide for a future detailed analysis.

The basic requirement of PHA is to form an ad hoc team of experienced individuals having familiarity with the material, substance, equipment, and the like. It is emphasized that experience and related expertise of team members are critical factors in performing a preliminary review. Each of these team members is asked to examine the occurrence of hazards in the area of their specialty and, as a team, to play devil's advocate.

6.5.2 HAZARD AND OPERABILITY REVIEW

HAZOP was originally developed for use in the chemical industry. HAZOP helps to identify problems prior to availability of full data concerning a product/system. The approach calls for the formation of a team made up of knowledgeable members with varying backgrounds, and in turn, the team brainstorms about potential hazards. An experienced individual chairs the team and serves as a facilitator during brainstorming sessions. The following basic steps are associated with the HAZOP technique [4, 10–11]:

- Choose the system/product to be analyzed.
- Establish a team made up of appropriate members.
- Describe the HAZOP process to all individuals forming the team.
- Set goals and time schedules.
- Conduct brainstorming sessions as appropriate.
- Document end results.

One major drawback of the HAZOP approach is that it does not take into consideration human error in the final equation.

6.5.3 TECHNIQUE OF OPERATIONS REVIEW

Just like in the case of HAZOP, TOR seeks to identify systemic causes rather than assigning blames. TOR allows management and workers to work jointly to perform analysis of workplace accidents, incidents, and failures [12].

This safety analysis technique is not new; it was developed by D. A. Weaver of the American Society of Safety Engineers (ASSE) in the early 1970s [4]. However, for almost 2 decades, there was no readily available user documentation on the technique. Consequently, its wide-scale application did not happen until the early 1990s because information on the technique did not become available at a larger scale.

TOR, basically, is a hands-on analytical methodology developed to determine the root system/product causes of an operation malfunction. The technique makes use of a worksheet containing simple terms requiring yes/no decisions. The basis for activating TOR is an incident occurring at a certain time and place involving certain individuals. The strength of the technique comes from involving line personnel in the analysis, and its weakness stems from the fact that it is designed as an after-the-fact process. The following steps are associated with the method:

- Form the TOR team containing appropriate members.

- Conduct a roundtable session for the purpose of departing common knowledge to all team members.
- Highlight one crucial factor that played an important role in the occurrence of accident/incident. This factor must be the result of team consensus and serves as an initial point to further investigations.
- Use the team consensus to respond to a sequence of yes/no options.
- Carefully evaluate the identified factors and ensure the existence of consensus among the team members.
- Prioritize the contributing factors.
- Establish corrective/preventive strategies with respect to each contributing factor.
- Implement strategies.

6.5.4 RISK ANALYSIS

Risk exists in all human activities that can either be health and safety related or economic related, for example, loss of equipment and production due to accidents involving fires, explosions, and so on. Risk analysis is an analytical tool that can be used to analyze the workplace, identify hazards, and develop mechanisms to eliminate such hazards. The fundamental reasoning behind risk analysis is that risk is lowered by decreasing the frequency and severity of the occurrence of hazard-related events.

There are basically two questions focused on by the risk analysis process:

- How often does a given event occur?
- What is the severity of the consequence of a given event?

The risk analysis process itself is composed of six steps: (1) scope definition, (2) hazard identification, (3) risk estimation, (4) documentation, (5) verification, and (6) analysis update. Over the years there have been many approaches developed to perform risk analysis [10, 11, 13, 14]. In the selection of such methods for use, factors such as appropriateness to the system/product/situation and scientific defensibility and simplicity of use must be carefully considered. The risk analysis methods may be grouped into two categories: hazard identification and risk estimation. The hazard identification approaches are concerned with the identification of inherent hazards and their types by taking advantage of historical data and experience gained by performing other risk analysis studies. Examples of the techniques belonging to the hazard identification category are HAZOP, event tree analysis (ETA), FTA, and FMEA.

The risk estimation approaches are concerned with the risk quantitive analysis requiring estimates of the frequency and consequences of hazardous events,

TABLE 6.1. Ratings of Hazards and Their Corresponding Frequency of Occurrence.

Rating	Rating Description	Frequency of Occurrence (per day)
1	Impossible	$>10^{-8}$
2	Extremely unlikely	$>10^{-6}$
3	Remote	10^{-5}
4	Occasional	10^{-4}
5	Reasonably probable	10^{-3}
6	Frequent	10^{-2}

system failure, and so on. Two of these methods are known as frequency analysis (FA) and consequence analysis (CA). One of the most effective risk analysis methods is described in Reference [15].

The method uses the ratings of hazards and their corresponding frequency of occurrence as given in Table 6.1. Obviously, as indicated in Table 6.1 the lowest rating of 1 means it is impossible that a given error/accident/incident will occur. On the other hand, the highest rating of 6 means that it is very likely that a given error/accident/incident will happen. Similarly, the severity levels of incidents/accidents/errors can also be rated as given in Table 6.2.

Some of the benefits of performing risk analyses are the identification of potential hazards and failure modes, improvement in the understanding of the system, making it possible to compare risks of similar devices/systems/equipment, and making better decisions concerning safety improvement expenditures.

6.5.5 FAILURE MODE AND EFFECTS ANALYSIS

FMEA is an important tool to evaluate system design from the reliability and safety angle. Originally, the method was developed in the early 1950s to evaluate the design of flight control systems [16]. The FMEA approach demands listing

TABLE 6.2. Severity Level Ratings of Incidents/Accidents/Errors.

Rating	Rating Description
1	Incidents/accidents/errors are not likely to cause an injury/damage property.
2	Incidents/accidents/errors may cause minor injury/minor occupational illness/minor damage.
3	Incidents/accidents/errors may cause severe injury/major loss.
4	Incidents/accidents/errors are almost certain to cause death/serious property damage.

potential failure modes of each and every component on paper and their effects on the listed subsystems, system, and the surroundings. The following steps are associated with FMEA [17]:

- Establish analysis scope.
- Collect data.
- List all possible failure modes, the identification, and description of each component.
- Assign failure rate/probability to each identified failure mode.
- List each failure mode effect or effects on subsystem, system, and so on.
- Enter appropriate remarks for each failure mode.
- Review each critical failure mode and take appropriate corrective measures.

A comprehensive list of publications on FMEA is given in Reference [18].

6.5.6 FAULT TREE ANALYSIS

FTA is a useful tool that can be employed to predict and prevent accidents. The method was originally developed at the Bell Telephone Laboratories to analyze the Minuteman Launch Control System with respect to reliability and safety in the early 1960s.

FTA may be described as an analytical methodology that uses graphic symbols to visually display the analysis process. Fault tree analysis begins by identifying an undesirable event, known as the top event, associated with a system. The events that could cause the occurrence of the top event are generated and connected by logic operators such as AND and OR. Thus, the fault tree itself is the logic structure relating the top event to the basic or the primary events. The basic principle underlying the construction of a fault tree is successively asking the question, "How could this event occur?"

Four basic symbols used in the construction of a fault tree are given in Figure 6.1. Each of the four symbols given in Figure 6.1 is described below:

- AND gate: This denotes that an output fault event occurs if all the input fault events occur.
- OR gate: This denotes that an output fault event occurs if any one or more of the input fault events occur.
- circle: This denotes a basic fault event or the failure of an elementary component.

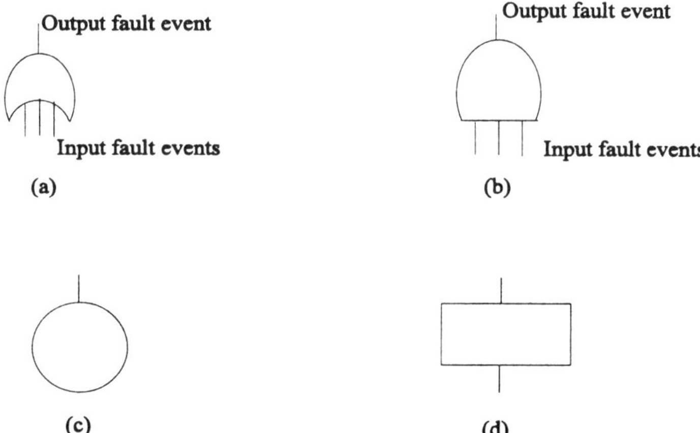

Figure 6.1 Basic fault tree symbols: (a) OR gate, (b) AND gate, (c) basic event (circle), (d) resultant event (rectangle).

- rectangle: This denotes a fault event that results from the combination of fault events through the input of a logic gate.

A safety-related fault tree may be developed through following four steps [4]:

- Decide on the top event accident/incident to be investigated.
- Identify the non-top event accidents/incidents/faults that may cause the top event to occur and assign appropriate symbols.
- Move downward through successive steps until basic fault/accident/incident events are identified.
- Review the end fault tree and make appropriate recommendations.

A detailed description of FTA is provided in Reference [17].

Example 6.1.

Develop a fault tree for the occurrence of a robot accident involving a human, which is caused by sudden robot movement. Figure 6.2 shows the fault tree for Example 6.1.

6.6 SAFETY COSTING

Soaring cost is an important factor in increasing attention on safety. For example, the U.S. Department of Health and Human Services reported that employers spent close to $22 billion to insure or self-insure against job-related

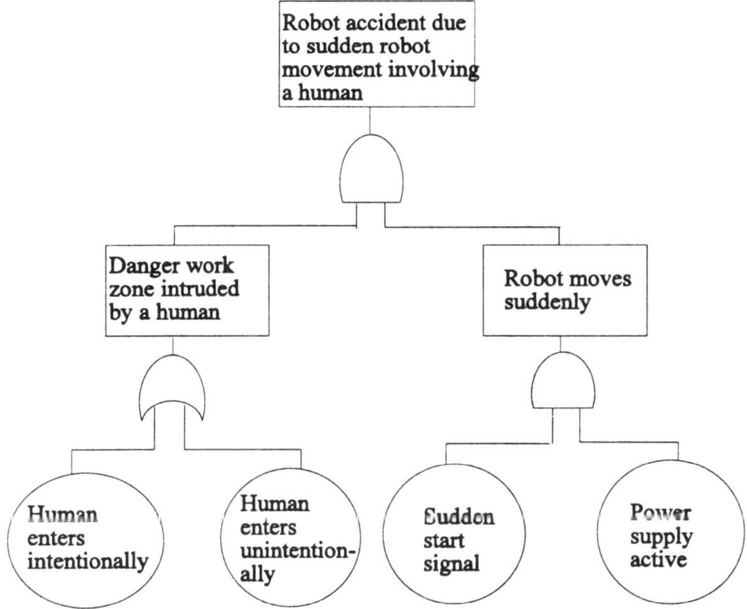

Figure 6.2 Fault tree for the top event: robot accident due to sudden robot movement involving a human.

injuries in 1980, while medical costs totalled $3.9 billion, along with $9.5 billion for compensation payments [19]. Nevertheless, according to a more recent estimate [4], the overall cost of accidents in the United States in a typical year is around $150 billion. This includes lost wages ($37.7 billion), medical expenses ($23.7 billion), motor vehicle property damage ($26.8 billion), insurance administration ($28.4 billion), indirect costs of work accidents ($22.5 billion), and fire-related losses ($9.4 billion).

There is a variety of possible resource losses (directly or indirectly involving safety and health and ultimately the cost) that should concern management and its sources: people, material, equipment, power/facilities, and premises/space [20].

The people resource losses include poor selection and recruiting, poor work methods, using wrong individuals for a task, absenteeism and lateness, people idle or slowed by accidents, and machine breakage that causes idle labor. Some of the elements of material resource losses are accidental damage during process, poor materials control, deterioration caused by poor storage, accidental damage from transport, and ineffective control of spares.

The equipment resource loss includes unsuitable tools and equipment, overloading equipment, lack of routine maintenance, no preventive maintenance, and poor equipment storage practices. The elements of the power/facilities resource loss are the wrong type of power connectors, insufficient light switches, overloaded electrical circuits, leaking air lines, no roof insulation, and more.

The premises/space resource loss includes poor plant layout, premises unsuitable for work, too small a work area, inadequate storage racks or bins, and inadequate materials handling equipment to use vertical space. Over the years people working in safety areas have developed several safety-related cost estimation models or methods.

6.6.1 PRODUCT SAFETY COST ESTIMATION MODEL

This model is proposed to estimate cost of safety for a product over its life span. Thus, the cost of safety for a product over its life cycle is expressed by [21]:

$$\text{PSC} = \text{APPC} + \text{IC} + \text{RC} + \text{ACL} - \text{R} \tag{6.1}$$

where

PSC = the cost of safety for a product over its life cycle
APPC = the accident prevention program cost
IC = the insurance cost
RC = the recall cost
ACL = the accident and claim loss
R = the reimbursements

6.6.2 HEINRICH METHOD

This method is named after H. W. Heinrich [5] who, over a half century ago, argued that, for every dollar of insured cost paid for accidents, there were four dollars of uninsured costs borne by the establishment [22]. Consequently, Heinrich expressed the indirect cost of an accident as follows:

$$IC_a = \sum_{i=1}^{11} C_i \tag{6.2}$$

where

IC_a = the total indirect cost associated with an accident
C_1 = the cost associated with the lost time of the injured employee
C_2 = the cost associated with the lost time by management
C_3 = the cost associated with the employees who stop work to watch or are involved in the action
C_4 = the cost associated with damage to machine/material/tools

C_5 = the cost associated with excitement or weakened morale
C_6 = the cost associated with the continuing payment of wages to employees
C_7 = the cost associated with welfare and benefits systems
C_8 = the cost associated with interference with production, lost orders, and so on
C_9 = the cost associated with lost time on the case by first aid and hospital workers not covered by insurance
C_{10} = the cost associated with the loss of profit on employee's productivity due to idle machines
C_{11} = the cost associated with overhead for the injured employee while in injured status

6.6.3 ACCIDENT PREVENTION COST ESTIMATION MODEL

The accident prevention program is an important element in an overall safety program. Thus, the cost associated with this program is expressed by [21]:

$$\text{APPC} = \text{SIP} + \text{RCA} + \text{ACA} - \text{RR} \tag{6.3}$$

where

APPC = the cost associated with the accident prevention program
SIP = the savings in insurance premiums
RCA = the recall cost that can be avoided
ACA = the accident and claim cost that can be avoided
RR = the reduction in reimbursements

6.6.4 SIMONDS METHOD

This method was developed by Professor R. H. Simonds of Michigan State College working in conjunction with the National Safety Council [4]. Simonds reasoned that the cost associated with an accident can be divided into two categories: insured and uninsured costs [23]. The estimation of insured cost is relatively simple in comparison to the determination of the uninsured cost. Thus, in order to estimate total uninsured cost, Simonds proposed dividing accidents into four distinct classes and using the following relationship [22, 23]:

$$\text{TUC} = (XC)a_1 + (YC)a_2 + (ZC)a_3 + (FC)a_4 \tag{6.4}$$

where

TUC = the total uninsured cost of accidents
a_1 = the number of workdays lost due to Class I accidents leading to permanent partial disabilities and temporary total disabilities

a_2 = the number of physician's cases associated with Class II accidents, i.e., the Occupational Safety and Health Act (OSHA) non-lost work-day cases that are attended by a doctor
a_3 = the number of first-aid cases associated with Class III accidents, i.e., those accidents in which first-aid was provided locally
a_4 = the number of noninjury cases associated with Class IV accidents, i.e., those accidents causing minor injuries that do not require the attention of a medical person

XC, YC, ZC, and FC are the average uninsured costs associated with class I, II, III, and IV accidents, respectively.

6.7 PRODUCT LIABILITY

This is one of the fastest growing areas of the law, and the United States has become the most litigious country in the world and has approximately 70% of the world's lawyers [4]. Until 1960, manufacturers were not held liable unless they produced flagrantly dangerous products. Today, the companies manufacturing various types of products are increasingly being sued by the product users, misusers, and even by abusers.

The following four landmark cases have established the basis for product liability suits in the United States [4]:

- *Henningson v. Bloomfield Motors, Inc.:* This case ruled in 1960 established the concept of breach of warranty.
- *Greenman v. Yuba Products, Inc.:* This case ruled in 1964 established the concept of strict liability in tort and negligent design.
- *MacPherson v. Buick Motor Company:* This case established the concept of negligent manufacture.
- *Van der Mark v. Ford Motor Company:* This case ruled in 1964 confirmed the concept established in the above *Greenman v. Yuba Products, Inc.* case, that is, negligent design.

The basis for negligence in engineering design could be any one of the following three areas [2, 24]:

- The design does not comply with accepted standards or the specified materials were of unsatisfactory strength.
- The manufacturer failed to provide required safety devices as part of the design.
- Concealment of danger is created by the design in question.

Also, it should be noted that approximately 60% of the product liability cases involved failure to provide satisfactory danger warning labels. Nevertheless, the common causes of product liability exposure include inadequate research during product development, faulty manufacturing, poorly written instructions, faulty product design, poor quality control, insufficient or poorly written warnings, and inadequate testing of product prototypes.

Many organizations usually develop a product safety management (PSM) program to minimize their exposure to product litigation and related problems. The success of such a program depends on its effective functioning. In any case, the program, at a minimum, should have three elements: coordinator, committee, and auditor.

6.7.1 COORDINATOR

Because the PSM program usually involves many departments (e.g., design, manufacturing, accounting, marketing, and service), the appointment of a PSM coordinator is essential. The key to the success of the PSM program coordinator is the level of his/her authority. Usually, it is considered that the higher the level of authority, the greater the chance of the program success. Nonetheless, the PSM coordinator should possess authority to undertake actions, including performing analysis of complaints, incidents, or accidents; coordinating all program documents; providing assistance in developing PSM program policy; making recommendations concerning product recall, field modification, redesign, and special analysis; performing PSM program audits; facilitating communication among all concerned individuals; developing a database concerning product safety and liability information; and liaising with agencies concerned with product safety and liability [23].

6.7.2 COMMITTEE

Many companies form a PSM committee because PSM is the responsibility of virtually all departments. Usually, the members of the committee belong to all major departments. Some of the advantages of having a PSM committee are that it provides a useful base for the coordinator to seek input on various issues, an efficient mode of communication with departments, and so on.

6.7.3 AUDITOR

Auditing is a vital element of the PSM program because it provides a feedback on the program effectiveness. The main function of the auditor is to determine the adequacy of the PSM program activities with respect to actual and potential exposures. In addition, the PSM auditor's specific duties include the identification of evidence concerning management's lack of commitment, bringing deficiencies to the management attention and making appropriate recommendations, monitoring the action taken by the management after the identification of a deficiency, and reviewing documentation associated with measures taken to rectify identified deficiencies.

6.8 PROBLEMS

(1) Write an essay on historical developments of the safety field.

(2) Describe the following classifications of hazard:
- electrical hazards
- energy hazards
- kinematic hazards
- environmental hazards

(3) Discuss the following types of injury:
- shearing-related injuries
- cutting and tearing-related injuries
- puncture-related injuries
- straining and spraining-related injuries

(4) Describe the safety measures during the following phases of a system:
- concept phase
- deployment phase
- definition phase

(5) Describe hazard and operability review (HAZOP) method.

(6) What is technique of operations review? Describe it in detail.

(7) List at least four methods used in risk analysis.

(8) Describe the Heinrich cost estimation method. Comment on its advantages.

(9) Describe the following two landmark cases that have helped to establish the basis of product liability law in the United States:
- *MacPherson v. Buick Motor Company*
- *Henningson v. Bloomfield Motors, Inc.*

(10) Discuss the factors that may form the basis for negligence in engineering design.

6.9 REFERENCES

1. Pliny the Elder (AD 23-79), *Natural History*, translated by H. Rackham, Harvard University Press, Cambridge, Massachusetts, 1947.

2. Dhillon, B.S., *Engineering Design: A Modern Approach*, Richard D. Irwin, Inc., Chicago, 1996.
3. Gloss, D.S., Wardle, M.G., *Introduction to Safety Engineering*, John Wiley & Sons, New York, 1984.
4. Goetsch, D.L., *Occupational Safety and Health*, Prentice-Hall, Inc., Englewood Cliffs, New Jersey, 1996.
5. Heinrich, H.W., *Industrial Accident Prevention*, 4th ed., McGraw-Hill, New York, 1959.
6. Roland, H.E., Moriarty, B., *System Safety Engineering and Management*, John Wiley & Sons, New York, 1983.
7. Dhillon, B.S., *Reliability Engineering in Systems Design and Operation*, Van Nostrand Reinhold Company, New York, 1983, Chapter 7 references.
8. DeReamer, R., *Modern Safety and Health Technology*, John Wiley & Sons, Inc., New York, 1980.
9. Hunter, T.A., *Engineering Design for Safety*, McGraw-Hill, Inc., New York, 1992.
10. *Risk Analysis Requirements and Guidelines*, CAN/CSA-Q634-91, prepared by the Canadian Standards Association, 1991. Available from Canadian Standards Association, 178 Rexdale Boulevard, Rexdale, Ontario, Canada, M9W 1R3.
11. Dhillon, B.S., Rayapati, S.N., Chemical Systems Reliability: A Survey, *IEEE Trans. on Reliability*, Vol. 37, 1988, pp. 199–208.
12. Hallock, R.G., Technic of Operations Review Analysis: Determine Cause of Accident/Incident, *Safety and Health*, Vol. 60, August 1991, pp. 38, 39, 46.
13. Vesely, W.E., *Engineering Risk Analysis in Technological Risk Assessment*, edited by Rice, P.F., Sagan, L.A., Whipple, C.G., Martinus Nijhoff Publishers, The Hague, 1984, pp. 49–84.
14. Covello, V., Merkhofer, M., *Risk Assessment and Risk Assessment Methods: The State-of-the-Art*, NSF report, 1984, National Science Foundation (NSF), Washington, D.C.
15. Chapanis, A., To Err Is Human, to Forgive, Design, *Proceedings of the ASSE Annual Professional Development Conference*, 1986, pp. 6–10.
16. Countinho, J.S., Failure Effect Analysis, *Transactions of the New York Academy of Sciences*, Vol. 26, 1964, pp. 564–584.
17. Dhillon, B.S., Singh, C., *Engineering Reliability: New Techniques and Applications*, John Wiley & Sons, New York, 1981.
18. Dhillon, B.S., Failure Mode and Effects Analysis: Bibliography, *Microelectronics and Reliability*, Vol. 32, 1992, pp. 719–732.
19. Lancianese, F., The Soaring Costs of Industrial Accidents, *Occupational Hazards*, August 1983, pp. 30–35.
20. Davis, F.E., An Australian Looks at Loss Control, *Professional Safety*, October 1983, pp. 40–41.
21. Hammer, W., *Product Safety Management and Engineering*, Prentice-Hall, Inc., Englewood Cliffs, New Jersey, 1980.
22. Raouf, A., Dhillon, B.S., *Safety Assessment: A Quantitative Approach*, Lewis Publishers, Boca Raton, Florida, 1994.
23. *Accident Prevention Manual for Industrial Operations: Administration and Programs*, 9th Edition, Prepared by the National Safety Council, Chicago, 1988.
24. Dieter, G., *Engineering Design*, McGraw-Hill Book Company, New York, 1983.

CHAPTER 7

Human Factors

7.1 INTRODUCTION

THE discipline of human factors exists because humans make errors in using machines; otherwise, it would be difficult to justify the discipline's existence. Since the terms *human factors*, *ergonomics*, *human factors engineering*, and *human engineering* have appeared interchangeably in the published literature, here human engineering is defined: a discipline that deals with designing manmade objects (machines/equipments) so that involved individuals can use them safely and in an effective manner to create environments ideal for the living and working of human beings [1]. Usually, the term *ergonomics* is less often used in North America than the other three terms. It is derived from two Greek words: *ergon* (meaning work) and *nomos* (meaning law).

The modern history of human factors may be traced back to Frederick W. Taylor who in 1898 performed studies to determine the most suitable design of shovels [2]. In 1911, Frank B. Gilbreth, a follower of Taylor's work, studied bricklaying, and that led to the invention of a scaffold [3, 4]. His new invention allowed bricklayers to perform their task at the most appropriate level at all times because the scaffold could be raised or lowered with ease quite quickly. Furthermore, the number of bricks laid per hour almost tripled (i.e., 120–350 bricks per hour), and the number of motions involved in laying a brick were reduced quite dramatically (i.e., from 18 to 5). In 1918, the U.S. government established laboratories at the Brooks Air Force Base and the Wright-Patterson Air Force Base to conduct research on various human factor–related areas [5].

In the late 1920s and early 1930s Hawthorne studies were basically directed to determine the effects of environment on worker output and ultimately resulted in finding the relationship between worker attitude and morale and output [6]. During World War II, it was realized that some military equipment increased significantly in sophistication and complexity, thus exceeding the capabilities of individuals to operate it. Consequently, experiment psychologists were enlisted to work jointly with engineers in designing various types of military equipment, such as radar consoles, combat information centers, and aircraft cockpits [7].

By 1945 human factors engineering was recognized as a specialized discipline. In 1954, Hertzberg et al. developed a database on body dimensions using United States Air Force (USAF) personnel as subjects [8].

In 1972, the Department of Defense released a document concerning human factors [9]. This military specification document outlined requirements for contractors developing equipment to be used by the services. The ongoing space exploration effort has also helped to increase the importance of human factors in equipment design. Currently, there are hundreds of published documents available on the subject in the form of textbooks, articles, and technical reports. Several research journals and annual conferences around the world are devoted to the subject. This chapter discusses various aspects of human factors with a special consideration to engineering design.

7.2 TYPES OF MAN-MACHINE SYSTEMS AND HUMAN ERROR RELATED SYSTEM FAILURES

Before we discuss the types of man-machine systems, let us first examine the difference between closed- and open-loop systems. A closed-loop system may simply be described as a continuous system performing some process requiring continuous control and feedback for its successful operation. In contrast, an open-loop system, when in action, requires no further control or at least cannot be further controlled. Although the distinctions between systems with respect to the above descriptions are not absolutely clear, the man-machine systems may be classified into the following three basic categories [10]:

- manual systems: These systems are composed of hand tools and other aids coupled with the human operator controlling the operation. The human operators use their own physical energy as a power source, as well as transmit to and receive from such tools a large amount of information.
- mechanical or semiautomatic systems: These systems are composed of well-integrated physical parts, such as different types of powered machine tools and are normally designed to carry out their assigned functions with minor variation. The machine typically provides power, and the human operator essentially performs the function of controlling.
- automated systems: These systems perform all operational functions, including sensing, processing, decision making, and action. The majority of such systems are of the closed-loop type, and usually the primary human functions involved with them are monitoring, programming, and maintenance.

Each year, hundreds of systems fail because of human error; three such examples are as follows:

- **Three Mile Island Nuclear Power Plant accident:** This accident occurred on March 28, 1979, at the number-two reactor. The subsequent investigation of the accident indicated that the basic cause of this mishap was an interwoven chain of mechanical, human, and institutional malfunctions [11].
- **Chernobyl Nuclear Power Plant accident:** This accident happened on April 26, 1986, at the number-four reactor. Two of the causes of the accident were attributed to problems with the systems analysis and deficiencies in the training of the plant's operator [12].
- **large stacker forklift truck accident:** This fatal accident occurred in Arizona during the transport of concrete sewer pipe sections. The subsequent investigation of the accident revealed that, since the control toggle switch design was contrary to normal human behavior, the forklift truck operator unintentionally pulled the toggle switch backward, thus causing the fatal accident [13].

7.3 HUMAN FACTORS CONSIDERATIONS IN PRODUCT DESIGN

In order to have an effective human compatible product/system, the relevant human factors must carefully be considered during the design stage. In addition, the major objectives should be to design a system/product that allows humans to perform in the most suitable manner, that possesses adaptability to humans, and that does not subject humans to extreme physical or mental stress or to hazards.

During the following stages of product/system design, the design engineer or others should consider human factors from various perspectives [14]:

- preconceptual stage: At this stage the design engineer or others should systematically define the mission and operational requirement, the functions required to perform each mission event, the performance requirements for each function, and the allocation of functions to hardware, software, or human components.
- conceptual stage: At this stage, in addition to the preconceptual stage tasks, the design engineer or others should also include analyses to define the most suitable design approach for accomplishing each hardware functional assignment; preliminary task descriptions of operators, maintainers, and users; preliminary definition of manning and training requirements; and so on. More specifically at this stage, the principal considerations should include the assigning of human roles in system

operation; deciding where, when, and how humans will interact with system elements; deciding in providing the best human work environments; and determining human constraint impact on the system and its associated elements. During system/product conceptualization, potential people categories and potential user characteristics must be carefully considered in detail as applicable. The potential people categories include operators, service personnel, production people, visitors, security people, training personnel, delivery personnel, technical and maintenance personnel, customers, and disaster control personnel. Similarly, the potential user characteristics include size, age, sex, motor response, strength, fatigue limits, intelligence level, physiological tolerance, cultural background, training and experience, reaction time, sensory response, coordination, adaptive limits, and cognitive response.

- predesign stage: At this stage the design engineer or others, in addition to reviewing the previous analyses, should also consider performing man-machine mockup and simulation studies, time line and link analyses, refined task analysis and so on.
- detailed design stage: At this stage the design engineer or others should at least consider creation of product/system statement, developing function-flow schematic diagrams, performing link analyses for all important human–equipment interfaces, identifying critical skill requirement specifications, creating and evaluating all critical man-machine mockups, and so on.

7.4 HUMAN BEHAVIORAL EXPECTANCIES

Humans have built-in tendencies towards certain items, as well as their behavioral/motor development varying with age. Such factors must be carefully considered by engineers during the design and development phase in order to produce successful end products. This section discusses human behavioral change from childhood to adulthood, human behavioral variation with age toward safety, and general human behavioral expectancies separately [14].

7.4.1 BEHAVIORAL CHANGE FROM CHILDHOOD TO ADULTHOOD

Over the years researchers around the world have established that human behavior changes with age. Thus, Table 7.1 presents selected examples of behavioral/motor development for ages 2 years to 21 years.

TABLE 7.1. Variations in Human Behavioral/Motor Development with Age.

Item No.	Age (years)	Description of Behavioral/Motor Development
1	2	Runs, can play simple catch and toss with ball, tends to use one hand more than the other, can operate a kiddie car around a chair, can string beads, can use scissors, etc.
2	5	Can choose medium-sized objects from a glass containing small, medium-sized, and large objects; can draw prism from copy; can differentiate between 3- and 15-g weights when their size is same; etc.
3	6–12	Can ride a bicycle, swim, roller-skate, etc.; can handle all garment fasteners in any way; handedness is established; the meanings of ounces, pounds, inches, feet, yards, and even miles are learned; etc.
4	17–21	Mature levels of behavior normally are completed (gradually), complicated skills can be acquired during this period, etc.

7.4.2 BEHAVIORAL VARIATION WITH AGE TOWARD SAFETY

Safety is a crucial factor in engineering design. Past research and experience are indicative of the fact that human attitudes toward safety vary with age. Thus, Table 7.2 presents the description of behavior toward safety for different age groups.

7.4.3 GENERAL BEHAVIORS

Over the years researchers have documented many different types of human behavior. The knowledge of such behaviors will be useful to engineering designers, especially when making design safety-related decisions. Table 7.3 presents selected such behaviors along with proposed design considerations [14].

7.5 COMPARISON OF HUMANS WITH MACHINES AND SELECTED METHODS FOR HUMAN FACTORS EVALUATION OF SYSTEM/PRODUCT/COMPONENT DESIGN

During the design phase, designers may have to make decisions whether to allocate certain functions to humans or to machines. In such situations the knowledge of humans' and machines' capabilities is absolutely essential to make rational decisions. Table 7.4 presents a comparison of humans' and machines' capabilities/limitations.

TABLE 7.2. Behavior Toward Safety for Different Age Groups.

Age Group No.	Age Group	Brief Behavioral Description
1	A: Infants and young children	They have curiosity about anything and everything. Unusual or new things motivate them to explore.
2	B: School-age children	They are tempted to take chances because of being accepted by the peer group and the importance of conforming. Consequently, often this age group of children suffers more serious accidents then those suffered by the younger group.
3	C: Teen-age children	They tend toward erratic behavior, resistance to authority, social antagonism, often daring behavior, etc.
4	D: Young adults	They tend to use their newly acquired strength, take more chances, attempt more daring and demanding feats, etc.
5	E: Early and middle years of adulthood	They tend to take only well-calculated risks, to be more oriented toward efficiency and cost effectiveness, and so on.
6	F. Older adults and persons close to their retirement age	They tend to take less risks and generally exercise greater caution.
7	G: Senior citizens	They tend to move more slowly and cautiously when engaging in all activities, lack patience, are unaware of obvious hazards, etc.

7.6 HUMAN SENSORY CAPABILITIES AND BODY MEASUREMENTS

Humans possess many useful sensors: touch, sight, hearing, smell, and taste. More specifically, man can sense vibration, acceleration (shock), temperature, pressure, position, rotation, and linear motion. A minute change in these sensations over a wide range is recognizable by humans, and then they automatically react to it. Some of these human sensory-related capabilities are discussed below [15, 16].

7.6.1 TOUCH

Touch is closely associated with a human's ability to interpret visual and auditory stimuli. Thus, in design work the touch sensor may be used to relieve eyes and ears of a part of the load.

TABLE 7.3. **General Behavioral Expectancies along with Proposed Design Considerations.**

Item No.	Expected Behavior or Habit	Proposed Design Consideration
1	Humans have become accustomed to specific color meanings.	Strictly observe current color-coding standards in products under design.
2	Humans often use their hands first to test or explore.	First of all, design product by paying a special attention to the handling aspect. Otherwise, make it explicity clear that its use needs a device supplied to remove the necessity for using hands.
3	Humans usually assume that an object is small enough to get hold of, as well as it being fairly light enough to pick up.	Incorporate some type of fastening in design to stop the object from being lifted. If not, provide a warning to potential users.
4	Humans, in general, regard manufactured products as being safe.	Design products so that they become impossible to be used improperly. If that is impractical, design a mechanism to make users think of possible hazards.
5	Human's attention is drawn to factors such as bright lights, bright and vivid colors, flashing lights, and loud noises.	Design in stimuli of satisfactory intensity when attention requires stimulation.
6	Humans get easily confused with unfamiliar things.	Avoid designing totally unfamiliar products to users.
7	Humans expect electrically powered switches to move upward, to the right, etc. to turn power on.	Design such devices per human expectations.
8	Humans, in the event of losing balance, instinctively reach for and grab the nearest thing.	Configure the design so that it includes adequate emergency supports.
9	Normally humans tend to hurry at one time or another.	Design the product so that it takes into consideration the element of hurry by humans involved.
10	Humans expect that faucets and valve handles rotate counterclockwise to increase the flow of a gas, liquid, or steam.	Design such devices per human expectations.
11	Usually, humans can get easily distracted by specific aspects of the product's features or by the recommended procedures for its application.	Review product design and use recommended procedures to eradicate distractions.
12	Usually, humans know very little about their physical limitations.	Learn about human basic characteristics and limitations and then develop appropriate design.

TABLE 7.4. A Comparison of Humans' and Machines' Capabilities/Limitations.

Item No.	Human	Machine
1	Is highly flexible with respect to task performance	Is relatively inflexible
2	Is highly capable of making inductive decisions in novel situations	Possesses little or no induction capability
3	Is limited in channel capacity	Can have channel capacity as much as required
4	Is prone to factors such as motion sickness, disorientation, and coriolis effects	Is independent of such effects
5	Is quite effective in recognizing and using information redundancy in real-world environments to simplify complex situations	Possesses rather limited perceptual constancy and is too expensive to have such a capability
6	Possesses a high degree of tolerance for factors such as ambiguity, uncertainty, and vagueness	Is very limited in tolerance with respect to such factors
7	Possesses an excellent memory	Is extremely expensive to have same capability
8	Is prone to stress as the results of interpersonal or other problems	Is independent of such problems
9	Is subject to social environments	Is free of social environments
10	Is not a good monitor of events that occur infrequently	Has an option to be designed to reliably detect events that do not occur frequently
11	Is subject to degradation of performance because of fatigue and boredom	Is not subject to such factors but its performance can degrade due to wear or lack of calibration
12	Is unsuitable for performing tasks such as data coding, transformation, or amplification	Is extremely suitable to perform such tasks
13	Possesses capability to interpret an input signal even under noisy, distractive, and similar conditions	Performs well only, usually, in ideal environment, i.e. clean, noise-free, etc.
14	Is capable of performing time-contingency analyses and predicting events in unfamiliar conditions	Is very poor at this aspect
15	Has relatively easy maintenance	Serious associated maintenance problems with increase in complexity
16	Performance efficiency affected because of the anxiety factor	Is independent of this problem
17	Is subject to physiological, psychological, and ecological needs	Is subject to only ecological needs

TABLE 7.4. Continued.

Item No.	Human	Machine
18	Possesses extremely limited short-term memory for factual matters	Can have as much short-term memory as desirable but subject to affordability
19	Is affected quite adversely by high g forces	Is independent of such forces
20	Is subject to departure from following an optimum strategy	Always executes the designed strategy
21	Possesses quite low absolute thresholds for sensing factors such as touch, vision, and audition	Can possess same capability but at an extremely high cost
22	Is capable of performing under transient overload and the performance degrades gracefully	Stops operating under overload conditions and normally fails all at once
23	Is capable of introducing errors through misidentification, reintegration, etc.	Makes use of such processes

The utilization of the touch sensor in technical work is not new; it has been used by craft workers for hundreds of years to detect irregularities and surface roughness in their work. Here, it is interesting to note from S. Lederman [17] that the detection accuracy of surface irregularities dramatically improves when the person moves an intermediate piece of paper or thin cloth over the object surface instead of when just the bare fingers are used. That is probably why the inspection of the coachwork of Aston Martin sports cars is performed by rubbing over the surface using cotton gloves.

In modern design work one typical example of the application of the touch sensor could be the recognition of control knob shapes with or without using other sensors.

7.6.2 SIGHT

Sight is stimulated by electromagnetic radiation of specific wavelengths, often referred to as the visible segment of the electromagnetic spectrum. The various parts of the spectrum (as seen by the eye) appear to vary in brightness. For example, during the daytime, the human eye is very sensitive to greenish-yellow light. Also, from different angles, the eye sees differently. Usually looking straight ahead, the eye can perceive all colors; however, the color perception decreases with the increase in the viewing angle. Additional factors concerning color are as follows [16]:

- inadequately illuminated areas or at night: In such situations, color makes little difference, and at a distance or for a small point source (e.g., small warning lights) it is impossible to distinguish between yellow, orange, blue, and green. In fact, all these colors appear to be white.
- color reversal: In the event of staring, for example, at a green or red light, and then glancing away, there is a possibility that the signal to the brain may reverse the color.

Some of the guiding factors for designers are as follows:

- Avoid placing too much reliance on color in the event of critical operations being performed by fatigued persons.
- Select colors so that color-weak personnel do not get confused.
- Try to use, whenever it is feasible, red filters with a wavelength greater than 6500 Angstrom units.

7.6.3 NOISE

Noise may simply be described as sounds that lack coherence. The reaction of humans to noise extends beyond the auditory system. It can lead to feelings such as boredom, fatigue, well-being, or irritability. Excessive noise may lead to various problems: reduction in the workers' efficiency, loss in hearing if exposed for long periods, adverse effects on work requiring a high degree of muscular coordination or intense concentration, and more. Two major physical characteristics are usually employed to describe noise: frequency and intensity. Usually, an ear can detect sounds of frequencies from 20–20,000 Hz, and it is most sensitive to frequencies between 600 and 900 Hz. It is important to note that, when people get exposed for long periods to noise of frequencies between 4000 and 6000 Hz, they normally suffer a major loss of hearing [15, 16].

Intensity is the other major characteristic and is usually measured in decibels (dB). A person exposed to a noise level of more than 80 dB may suffer temporary or permanent loss of hearing. Table 7.5 presents various noise intensity levels and their associated effects on humans and the noise source typical examples [10, 15].

Table 7.6 presents Occupational Safety and Health Administration (OSHA) permissible time duration versus noise level selected guidelines [10]. Table 7.7 presents the effects on humans of selected noise levels [14]. Additional information on design guidelines for noise may be found in References [10, 14, 18].

TABLE 7.5. **Noise Intensity Levels and Their Associated Noise Source Typical Examples and Effects on Humans.**

Item No.	Intensity Level in Decibels (dB)	Noise Source Typical Example	Effect on Humans
1	10	Motion picture sound studio	Acceptable
2	20	Voice whisper	Acceptable
3	40	Quiet residential area	Acceptable
4	56	Household ventilating fan	Acceptable
5	60	Normal conversation	Acceptable
6	70	Heavy traffic	Acceptable
7	76	"Quiet" factory area	Acceptable
8	90	City bus	Above this point, the reduction in efficiency may occur
9	110	Punch press	Unacceptable or dangerous
10	120	Loud thunder	Unacceptable or dangerous

TABLE 7.6. **Occupational Safety and Health Administration Permissible Time Duration versus Noise Level Selected Guidelines.**

Item No.	Noise/Sound Level (dB)	Exposure per Day (hours)
1	105	1
2	100	2
3	97	3
4	95	4
5	90	8

TABLE 7.7. **Effects on Humans of Selected Noise Levels.**

Noise/Sound Level in dBs	Duration Time	Spectrum/Typical Source Example	Anticipated Effect
100	8 hours (maximum)	—	Dramatic reduction in alertness. If no protection, temporary hearing loss in frequency region 600–1200 Hz
105	2 minutes	Jet engine	Stereoscopic acuity, reduction in visual acuity, etc.
110	8 hours	Machinery noise	Digestive disorders and chronic fatigue
120	1 hour	Broadband	Loss of equilibrium
150	2 minutes	1–100 Hz	Changes in respiratory rhythm, reduction in visual acuity, etc.

7.6.4 VIBRATION

Vibration is concerned with the effects on individuals' performance of periodically occurring mechanical related forces impinging on body tissues. The effect of vibrations on the human body greatly depends upon the impinging energy physical parameters: application direction, body tissue/organs/etc. absorption coefficient and mechanical impedance. However, it is interesting to note that the human body's reaction to vibration and resonating stimuli is very much the same as that of the mechanical system masses and springs.

There are many parameters of vibrations: amplitude, frequency, velocity, acceleration, and jolt. In fact, for a fixed frequency, velocity, acceleration, and jolt are successive derivatives of amplitude with respect to time. Past experience indicates that large amplitude and low frequency vibrations contribute to the following [16]:

- headaches
- eye strain
- motion sickness
- fatigue
- deterioration in ability to read and interpret instruments

Also, the low amplitude and high frequency vibrations can be fatiguing. Some of the guidelines/tips for reducing vibration and motion effects are as follows [16, 18]:

- Minimize or eliminate shock and vibrations through design and/or using devices such as shock absorbers, springs, and cushioned mountings.
- Vertical vibrations affect seated personnel most.
- To reduce vibration transmissions to a seated individual's body, use damping materials or cushioned seats. Also, eliminate vibrations of 3–4 Hz because this is a seated individual's vertical truck resonant frequency.
- Eliminate vibrations greater than 0.08 mil amplitude for crucial maintenance or other operations requiring letter or digit discrimination.

7.6.5 HUMAN BODY MEASUREMENTS

Since humans usually operate and maintain engineering products, their body measurement information is crucial to designers so that allocated tasks to humans can be performed as effectively as possible. Usually, this kind of

TABLE 7.8. **Selected Body Dimensions of the U.S. Adult Population (18 to 79 years).**

Item No.	Body-Related Description	5th Percentile (in inches)		95th Percentile (in inches)	
		Male	Female	Male	Female
1	Standing height	63.6	59	72.8	67.1
2	Seated length	17.3	17.0	21.6	21.0
3	Weight	126 (lb)	104 (lb)	217 (lb)	199 (lb)
4	Seated width	12.2	12.3	15.9	17.1
5	Sitting height	33.2	30.9	38.0	35.7
6	Seated eye height	28.7	27.4	33.5	31.0

information is required during the early phases of design to ensure that the product under design accommodates humans of various sizes and shapes. Most of the time human body related requirements are outlined in the design specification, especially when developing a piece of equipment for use in a military application. For example, MIL-STD-1472 [19] states, "Design shall insure operability and maintainability by at least 90 percent of the user population." It also states, "The design range shall include at least the 5th and 95th percentiles for design-critical body dimensions." In addition, the standard also requires that the use of anthropometric data should take into consideration the factors listed below:

- the body position during the task performance
- the nature and frequency of tasks to be performed
- difficulties associated with intended tasks
- increments in the design-critical dimensions imposed by protective garments, the need to compensate for obstacles, and so on
- the imposed mobility or flexibility requirements of the task

As for commercially designed engineering products for use by the population at large, Table 7.8 presents selected body-related dimensions of the U.S. adult population (18 to 79 years) [14, 19]. The selective body dimension deta for ground troops per MIL-STD-1472D [19, 21] is given in Table 7.9.

7.7 FORMULAS FOR DESIGN SPECIALISTS

Over the years researchers have developed various types of mathematical formulas to estimate human factors–related information. This section presents such selected formulas.

TABLE 7.9. The Selective Body Dimension Data for U.S. Ground Troops.

Item No.	Body Related Description	5th Percentile (in inches)	95th Percentile (in inches)
1	Height	64.1	73.1
2	Waist height	38.0	45.3
3	Kneecap height	18.7	23.1
4	Sitting height, relaxed	27.6	32.5
5	Weight	122.4 (lb)	201.9 (lb)
6	Neck circumference	13.5	16.1
7	Waist circumference	26.9	37.8
8	Hand length	6.85	8.13
9	Foot length	9.65	11.41
10	Head circumference	20.94	23.16
11	Shoulder (bideltoid) breadth	16.3	19.6

7.7.1 THE DECIBEL

The level of noise/sound intensity is measured in term of decibels. The bel (B), the basic unit of measurement, is named in honor of Alexander Graham Bell, the inventor of the telephone. Thus, the sound-pressure level (SPL), in decibels, is expressed as [10, 22]

$$\text{SPL(dB)} = 10\log_{10}\left(\frac{P^2}{P_0^2}\right) \tag{7.1}$$

or

$$\text{SPL(dB)} = 20\log_{10}\left(\frac{P}{P_0}\right) \tag{7.2}$$

where

P^2 = the sound pressure squared of the sound one desires to measure
P_0^2 = the standard reference sound pressure squared representing zero decibels. More specifically, P_0 is ordinarily the faintest 1000 Hz tone that an average young adult can hear.

7.7.2 ENERGY COST

In many instances, humans perform frequent or virtually continuous lifting tasks; under such a circumstance, the work efficiency (i.e., the energy cost per unit of work) is basically influenced by the lifting range and the work pace. The following formula is proposed by S. W. Frederick [23] to estimate the human energy cost in kilocalories for given lifts per hour, weight, and lift range:

$$\text{kcal/h} = n \cdot h \cdot w \cdot E/1000 \tag{7.3}$$

where

kcal/h = the kilocalories per hour
E = the energy consumption in lifting per unit of work for various weights and specified lift ranges (gram calories per ft-lb). The values of E may be obtained from References [1, 10].
n = the number lifts per hour
w = the weight in lb
h = the lifting height expressed in feet

7.7.3 CHARACTER HEIGHT—FORMULA I

For a comfortable arm reach to perform control and adjustment-oriented tasks, usually the instrument panels are located at a viewing distance of 28 inches (in.). Thus, the sizes of letters, markings, and numbers are based on this viewing distance. On occasion, the need may arise to vary the viewing distance; under such circumstances the required character height can be estimated by using the following relationship [1, 10]:

$$\text{NHC} = (\text{SHC})\,(\text{RD})/28 \tag{7.4}$$

where

RD = the required distance expressed in inches
NHC = the new estimated height of a character at RD expressed in inches
SHC = the standard/recommended character at a viewing distance of 28 in.

Example 7.1

Assume that a meter has to be read at 84 in. The standard/recommended numeral height at a viewing distance of 28 in. at low luminance is 0.30 in. Estimate the height of numerals for the 84-in. viewing distance.

Substituting the specified data into Equation (7.4), we get

$$\text{NHC} = (0.31)(84)/28 = 0.93 \text{ in.}$$

Thus, the new required height of numerals is 0.93 in.

7.7.4 CHARACTER HEIGHT—FORMULA II

Usually, the ability of humans to make visual discrimination is dependent on factors such as illumination, size, and exposure time. In 1959, Peters and Adams

[24] developed the following formula to estimate character height by taking into consideration factors such as viewing distance, illumination, importance of reading accuracy, and viewing conditions:

$$\mathrm{CH} = \alpha(\mathrm{VD}) + C_1 + C_2 \tag{7.5}$$

where

CH = the character height expressed in inches
α = a constant whose value is 0.0022
VD = the viewing distance expressed in inches
C_2 = the correction factor for importance. For important items such as emergency labels, its recommended value is 0.075 and for other items $C_2 = 0$
C_1 = the correction factor for illumination and viewing conditions. Its recommended values for different situations are 0.26 (below 1 footcandle, unfavorable reading conditions), 0.16 (below 1 footcandle, favorable reading conditions), 0.16 (above 1 footcandle, unfavorable reading conditions), and 0.06 (above 1 footcandle, favorable reading conditions).

Example 7.2
Assume that the viewing distance of an instrument panel is 36 in. Calculate the height of the label characters to be used at the panel for $C_1 = 0.16$, and $C_2 = 0.075$.

Substituting the given data into Equation (7.5), we obtain

$$\begin{aligned} CH &= (0.0022)(36) + 0.16 + 0.075 \\ &= 0.3142 \text{ in.} \end{aligned}$$

Thus, the estimated height of the labeling characters is 0.3142 in.

7.7.5 NOISE REDUCTION

In industrial or other settings, noise could be a major problem. The overall noise reduction is not only dependent on the transmission loss alone, but also on factors such as the total absorption properties of walls in the receiving room and the area of the wall transmitting sound. Consequently, the following formula can be used to estimate total noise reduction [1]:

$$\mathrm{TNR} = \theta + \log(\mathrm{AP}/\mathrm{AS}) \tag{7.6}$$

where

TNR = the total noise reduction
θ = the transmission loss of various types of materials of the varying thicknesses, expressed in decibels

AP = the total absorption properties of walls in the noise receiving room
AS = the area of wall transmitting sound expressed in ft^2

In turn, the θ is defined by

$$\theta = \mathrm{AS}/(\alpha_1 \mathrm{AS}_1 + \alpha_2 \mathrm{AS}_2 + \alpha_m \mathrm{AS}_m) \tag{7.7}$$

where

α_i = the ith transmission coefficient of material in question; $i = 1, 2, 3, m$
AS_i = the ith corresponding area of the material in question

7.7.6 REST PERIOD

Usually, when humans perform various types of tasks, rest periods are required. The length of the rest period may vary according to the task being performed. In design work (as applicable) the factor of rest period requirement has to be taken into consideration carefully for its ultimate effectiveness. The following formula, depending on task average energy cost, was developed by Murrell [25] to estimate scheduled or nonscheduled rest periods:

$$\mathrm{RP} = \mathrm{TWT}(\mathrm{AC} - \mathrm{SC})/(\mathrm{AC} - \lambda) \tag{7.8}$$

where

RP = the required rest expressed in minutes
TWT = the total working time expressed in minutes
AC = the average energy cost/expenditure expressed in kilocalories per minute of work
SC = the kilocalories per minute adopted as standard
λ = the approximate resting level expressed in kilocalories per minute with its value taken as 1.5

Average energy cost or expenditure in kilocalories per minute/heart rate in beats per minute, respectively, for various different tasks are as follows [1, 26]:

- 8/150 (unload coal cars in power plants)
- 4.5/112 (cleaning floors and tables)
- 3.7/113 (packing on conveyor)
- 2.5/113 (bagging and packing paper rolls)

Example 7.3

Assume that a person is performing a task for 120 minutes and his/her average energy expenditure is 5 kilocalories per minute. Calculate the required rest period if SC = 4 kilocalories per minute.

Substituting the given data into Equation (7.8) yields

$$\begin{aligned} RP &= (120)(5 - 4)/(5 - 1.5) \\ &= 37.1 \text{ minutes} \end{aligned}$$

Thus, the required rest period is 37.1 minutes.

7.8 HUMAN FACTORS GENERAL GUIDELINES FOR PRODUCT/SYSTEM DESIGN

Even though performing effective human factors analyses in product/system design requires in-depth knowledge of the subject, the general guidelines serve as a vehicle for such analyses. Some of those guidelines are as follows [14]:

- Review the product/system mission/purpose/objective with care, especially from the human factors aspect.
- Use the services of human factors specialists as applicable or practical.
- Develop a human factors design checklist for use in design and production cycle.
- Acquire applicable human factors design guide and reference documents.
- Review the entire initial design concept ideas thoroughly.
- Make use of the above checklist throughout the design and production cycle.
- Use mockups to "test" the effectiveness of all user–hardware interface designs.
- Conduct experiments when cited reference guides do not provide satisfactory information for design-related decisions.
- Fabricate a hardware prototype (if possible) and evaluate this under real-life environments.
- Critically review final production drawings with respect to human factors.
- Conduct field tests of the product design prior to approving it for final delivery to the customer/user.

7.9 PROBLEMS

(1) Define the following terms:
   - human factors
   - ergonomics
   - human engineering

(2) Describe three types of man-machine systems.
(3) Describe at least two well-publicized human error–related system failures.
(4) Discuss at least seven general human behavioral expectancies and associated design measures.
(5) Make a comparison of humans with machines.
(6) What are the human sensory capabilities? Describe each of them in detail.
(7) Define a decibel by solving one numerical example.
(8) Assume that an instrument panel has to be read at 60 in. and the standard recommended height at a viewing distance of 28 in. at low luminance is 0.25 in. Calculate the height of characters for the specified viewing distance.
(9) A person is to carry out a task for 180 minutes and his/her estimated average energy expenditure is 4 kilocalories per minute. Determine the required rest period if the adopted standard kilocalories per minute are 5.

7.10 REFERENCES

1. Dale Huchingson, R., *New Horizons for Human Factors in Design*, McGraw-Hill Book Company, New York, 1981.
2. Chapanis, A., *Man-Machine Engineering*, Wadsworth Publishing Company, Inc., Belmont, California, 1965.
3. Gilbreth, F.B., *Bricklaying System*, Published by Mryon C. Clark, New York, 1909.
4. Dhillon, B.S., *Human Reliability: With Human Factors*, Pergamon Press, Inc., New York, 1986.
5. Meister, D., Rabideau, G.F., *Human Factors Evaluation in System Development*, John Wiley & Sons, New York, 1965.
6. Gilmer, B.V.H., *Industrial and Organizational Psychology*, McGraw-Hill Book Company, New York, 1971.
7. Fitts, P.M., Engineering Psychology and Equipment Design, in *Handbook of Experimental Psychology*, edited by S.S. Stevens, John Wiley & Sons, New York, 1951, Chapter 35.
8. Hertzberg, H.T.E., et al. *Anthropometry of Flying Personnel—1950*, Technical Report 53-321 (USAF), Department of Defense, Washington, D.C, 1954.
9. MIL-H-46855, *Human Engineering Requirements for Military Systems, Equipment, and Facilities*, May 1972, Department of Defense, Washington, D.C.
10. McCormick, E.J., Sanders, M.S., *Human Factors in Engineering and Design*, McGraw-Hill Book Company, New York, 1982.
11. Torrey, L., The Accident at Three Mile Island, *New Scientist*, November 8, 1979, pp. 424–428.
12. Wilkie, T., Soviet Engineers Admit Failings in Reactor Design, *New Scientist*, August 28, 1986, pp. 14–15.
13. Ross, B., What Is a Design Defect? in *Structural Failure, Product Liability and Technical Insurance*, edited by Rossmanith, H.P., Elsevier Science Publishing Company, Inc., New York, 1984, pp. 44–46.

14. Woodson, W.E., *Human Factors Design Handbook*, McGraw-Hill Book Company, New York, 1981.
15. AMCP 706-133, *Engineering Design Handbook: Maintainability Engineering Theory and Practice*, Prepared by the United States Army Material Command, 5001 Eisenhower Ave., Alexandria, VA 22333, 1976.
16. AMCP 706-134, *Engineering Design Handbook: Maintainability Guide for Design*, Prepared by the United States Army Material Command, 5001 Eisenhower Ave., Alexandria, VA 22333, 1972.
17. Lederman, S., Heightening Tactile Impressions of Surface Texture, in *Active Touch*, edited by G. Gordon, Pergamon Press, Elmsford, New York, 1978.
18. Altman, J.W., et al., *Guidelines to Design of Mechanical Equipment for Maintainability*, United States Air Force (USAF), Aeronautical Systems Division, Dayton, Ohio, Technical Report ASD-TR-61-381, August 1961.
19. MIL-STD-1472 (-), *Human Engineering Design Criteria for Military Systems, Equipment, and Facilities*, Department of Defense, Washington, D.C.
20. Stoudt, H.W., et al., *Weight, Height, and Selected Body Dimensions of Adults, United States 1960–1962*, Public Service Publication No. 1000, Ser. 11, No. 8, Washington D.C.
21. Hunter, T.A., *Engineering Design for Safety*, McGraw-Hill, Inc., New York, 1992.
22. Adams, J.A., *Human Factors Engineering*, Macmillan Publishing Company, New York, 1989.
23. Frederick, S.W., Human Energy in Manual Lifting, *Modern Materials Handling*, Vol. 14, 1959, pp. 74–76.
24. Peters, G.A., Adams, B.B., Three Criteria for Readable Panel Markings, *Product Engineering*, Vol. 30, No. 21, 1959, pp. 55–57.
25. Murrell, K.F.H., *Human Performance in Industry*, Reinhold Publishing Company, New York, 1965.
26. Davis, H.L., Faulkner, W.T., Miller, C.L., Work Physiology, *Human Factors*, Vol. 11, April 1969.

CHAPTER 8

Total Quality Management

8.1 INTRODUCTION

THE global competition has greatly influenced the ways of doing present-day business. The age-old belief of many companies, that is, the customer is always right, has a new twist—customer satisfaction. Thus, for the matter of survival, companies are focusing on customer satisfaction. The usual tool used by these organizations for customer satisfaction is the total quality management (TQM) concept.

TQM may simply be described as a quality emphasis that encompasses the entire organization, from supplier of parts/services to customer. In turn, TQM stresses management commitment for an active company-wide drive toward excellence in all areas of products and services important to customers [1].

The origin of the TQM movement goes back to the late 1940s, with quality gurus such as Deming, Juran, and Feigenbaum [2]. Thus, the TQM concept was not born in Japan as is believed by many people but, instead, was tested and enriched in that country. After the defeat of Japan in World War II, General Douglas MacArthur took two American engineers, Sarasohn and Protzman, to Japan to teach the Japanese modern manufacturing plant management [3]. In fact, Sarasohn wrote a book in Japanese entitled *The Industrial Application of Statistical Quality Control*, that resulted in the Japanese Union of Scientists and Engineers (JUSE) asking General MacArthur to bring W. E. Deming to lecture in Japan [4].

In July 1950, Deming delivered his first lecture on Elementary Principles of Statistical Control of Quality to 230 Japanese engineers and scientists. One year later, in 1951, the Japanese Union of Scientists and Engineers established a prize named after Deming to be awarded to a company that demonstrated the most effective implementation of policies and measures concerning quality [3].

As a result of effective quality measures, Japan became a leader in the world marketplace by 1979.

Many American companies became quite alarmed with the emergence of Japan in its leadership role. For example, Xerox Corporation was surprised to discover in 1979 that the Japanese copiers were not only equal in quality but took half the time to bring to the marketplace at almost half the cost.

In 1985, the term *total quality management* was coined by an American behavioral scientist named Nancy Warren [5]. After witnessing the success of the Deming prize in Japan, the U.S. government established the Malcolm Baldrige Award in 1987 for organizations that demonstrated successful implementation of quality assurance measures and procedures. In 1988, the Malcolm Baldrige Award was given to the cellular telephone division of Motorola for its effort to reduce defects from 1000 per million to 100 per million during the period between 1985 and 1988 [6, 7]. Also, this division set a target to reduce defects to four defects per million in 1992.

This chapter presents various useful aspects of TQM.

8.2 TQM BASICS AND COMPARISON OF TRADITIONAL AND TOTAL QUALITY MANAGEMENT

Many times, there could be good scientific innovations, but because of the lack of a TQM management methodology, the companies/nations may lose the competitive advantage. For example, during the period from the mid-1950s to the mid-1980s, U.K. scientists won 26 Nobel Prizes for new discoveries versus only four for Japan [8]. Further, during the same time period, Japanese industry successfully took a major portion of the world market. To add insult to injury, people in western countries, especially in the United States, started to prefer products made in Japan, not just because of less cost, but because of better quality. In addition, Deming [9] pointed out that 85% of the quality-related problems in U.S. industry were the result of poor management. It means quality and good management goes in hand in hand. Thus, before we go any further let's examine the term *total quality management* in detail.

Obviously, this term is composed of three words, each of which is discussed below:

- total: This calls for the team effort of all parties (i.e., inside and outside) of the organization to satisfy customers. There are various factors involved in having a successful supplier–customer relationship: regular monitoring of suppliers' products and processes by the customer, customers making suppliers understand their needs in an effective manner, development of a customer–supplier relationship based on mutual trust and respect, and customers developing their in-house requirements.

TABLE 8.1. A Comparison between the Traditional Quality Assurance Program and TQM.

No.	Factor	Traditional Quality Assurance Program	Total Quality Management
1	Quality defined	Products satisfy specifications	Products suitable for consumer use/application
2	Definition	Product-driven	Customer-driven
3	Customer	Ambiguous comprehension of customer/consumer needs	Well-defined mechanism to understand and meet customer needs
4	Objective	Discover errors	Prevent errors
5	Cost	Better quality leading to higher cost	Better quality reduces cost and increases productivity
6	Decision making	Followed top-down approach	Followed team approach with team of employees
7.	Quality responsibility	Quality control department/ inspection center	Everyone in the organization is involved

- quality: This is deceptively simple but endlessly complicated. That is probably why there are numerous definitions of quality. In any case, quality must be seen from the customer perspective. This view is further reinforced by the result of a survey [10]: 82% of the respondents stated that quality is defined by the customer and not by the supplier.
- management: The approach to management is critical in determining company ability to attain corporate goals and to allocate resources effectively. The TQM concept calls for a dramatic change in involving employees in company decision making because their contribution and participation are crucial to mold all areas of business in providing quality products to customers. It means that the companies contemplating the adoption of the TQM concept will have to view their employees with new visions. Another important factor involved in improving product quality is the commitment of top management to show consistent enthusiasm so that the employees seriously consider the importance of quality.

There are many differences between traditional quality assurance management and TQM. Thus, for the effective practicing of the TQM concept, a proper understanding of such differences is essential. Table 8.1 presents the comparison of traditional and total quality management [3, 7].

8.3 TQM PRINCIPLES AND ELEMENTS

Broadly speaking, TQM is based on two fundamental principles: customer satisfaction and continuous improvement. Customers could either be internal or external, but the external customers are not part of the organization manufacturing the good/product or service. The use of the "market-in" concept allows a strong customer orientation that recognizes that every work process is made up of stages. In turn, during each stage, customer input is sought for determining the changes that should be made to better satisfy the customer's requirements [11].

Continuous improvement is an important factor to meet the quality challenge. In fact, Tom Peters [12] stated, "Excellent firms don't believe in excellence—only in constant improvement and constant change." The incremental improvement and maintenance tasks are accomplished through process improvement and control. In any case, with respect to TQM, management performs two functions: (1) continuously makes improvements to methods and procedures used currently through process control and (2) directs efforts to achieve important advances in concerned processes.

The important elements of TQM may be classified as follows [11]:

- management commitment and leadership role
- team effort
- training
- supplier participation
- cost of quality
- customer service
- statistical techniques and methods

Management commitment and a leadership role are quite crucial for the success of the TQM concept. TQM will thrive only when the top management is genuinely committed to its cause and establishes TQM as a top priority. This can only be achieved after a thorough understanding of the TQM approach by senior management. After this, the management can establish new goals and directions for the organization and then play the leadership role toward the realization of those goals and directions.

In order to successfully implement and continuously apply the TQM process, the formation of quality teams is quite essential. The ultimate objective of the team approach is to get all individuals involved in the TQM concept, including customers, subcontractors, and vendors. The key to the quality team organization is the advisory committee that is responsible for developing the team structure, in addition to establishing a mechanism for team formation and the implementation process. Usually, the committee membership is determined by the top management involved in TQM implementation. Even though the committee members normally have a good understanding of the TQM concept, usually the committee is supported by a group of consultants. These consultants assist the committee in areas such as training material development, TQM systems and structure development, TQM promotion among management and nonmanagement personnel, understanding of TQM tools and techniques, liaison between management and employees, and TQM implementation [13, 14].

The quality team membership is voluntary, and it may range from 3 to 15 members. Usually, the team meets once a week for about an hour. The team meetings are chaired by a team leader, usually belonging to management background (i.e., a supervisor or a manager). Such a person possesses characteristics such as group leadership skills, skill in group dynamics, communication skills, skill in statistical methods, and presentation skills. In turn, the team members either possess skills or are trained in areas such as statistics, cost-benefit analysis, planning and controlling projects, flow charting, brainstorming, presentation methods, and public relations.

Training is an important element in TQM; in fact, a Japanese axiom concerning it states that quality starts with training and ends with training [15]. Under TQM, quality becomes the responsibility of each individual employed in the organization, which means the training effort must be targeted for every hierarchy level of the firm in question. Furthermore, the training should be tailored accordingly for engineers, management, field labor, technicians, support personnel, and home and field staff.

A study of over 200 companies that implemented TQM revealed that the skills in areas such as leadership, human interaction, and initiative are quite critical for the success of any quality improvement effort [16]. In any case, the training program should cover areas such as fundamentals of TQM, quality measurement cost, team problem solving, cause-and-effect analysis, interpersonal communication and interaction, and rudimentary statistical techniques.

Supplier participation is an essential component of TQM because a company's ability to manufacture a quality product, to a large extent, depends upon the type of relationship among the parties involved in the process: the processor, the customer, and the supplier. The supplier responsiveness can only be achieved through partnership and mutual trust. Nowadays, some firms require their suppliers to have formal TQM programs as a condition for future business [17].

The cost of quality is the primary quality measurement tool [18, 19] and is used for purposes such as monitoring the TQM process effectiveness, choosing quality improvement projects, and justifying the cost to doubters. Reference [20] defines the cost of quality, C_q, as follows:

$$C_q = \mathrm{QMC} + \mathrm{DC} \tag{8.1}$$

where

$$\mathrm{QMC} \equiv \mathrm{PC} + \mathrm{AC} \tag{8.2}$$

QMC = the quality management cost
DC = the deviation cost
PC = the prevention cost
AC = the appraisal cost

Two clear benefits of the cost of quality are (1) its effectiveness to raise awareness about quality and (2) to communicate to management the advantages of the TQM concept in terms of dollars.

As the number of firms using the TQM concept increases, the demand for improved quality also increases simultaneously, and the customer involvement becomes increasingly important. This results from the fact that the TQM process calls for universal involvement. The application of the TQM concept to the customer area in the form of joint teams usually leads to customer satisfaction. The basic idea behind the formation of these teams is to measure the level of customer satisfaction through listening and communicating with customers. Ultimately, these teams develop joints goals, plans, and controls.

Statistical techniques and methods are useful problem-solving tools for the TQM concept. Examples of these tools are Pareto diagrams, control charts, cause-and-effect diagrams, graphs and histograms, scatter diagrams, and flowcharts [15, 21, 22]. Such tools are useful for various purposes, including identifying and separating quality problem causes; verifying, repeating, and reproducing measurements based on data; communicating the information in a language that can easily be understood by all team members; and making decisions on facts based on data rather than the opinions of individuals [11, 23].

8.4 DESIGNING FOR QUALITY AND DEMING'S FOURTEEN-POINT APPROACH TO TQM

Quality begins during the product design specification writing phase and continues to its operational phase. In any case, there are various elements of product design that can adversely affect quality. Some of the examples of such elements are poor part tolerences, parts difficult to fabricate repeatedly because of poor design features, use of fragile parts, and the lack of design refinement, resulting in the use of more parts than necessary to perform the desired functions. There are measures that may be taken to improve quality during the product design phase, including minimizing number of parts and part numbers, eliminating adjustments, designing for efficient and satisfactory testing, designing for robustness using Taguchi methods, eliminating engineering changes on released products, laying out parts for reliable process completion, simplifying assembly and making it foolproof, using repeatable and well-understood processes, selecting components that can withstand process operations, and laying out parts for reliable process completion [24, 25].

In any case, W. E. Deming has outlined a fourteen-point approach to TQM that should automatically help directly or indirectly to improve design quality. The points of the approach are listed below [3, 26, 27]:

- Develop consistency of purpose for improving services. The development of a mission statement is key to this point, and the mission statement should address issues such as quality philosophy, profit distribution, investors, customers, long-term corporate objectives, employees, and growth plans.

- Lead to promote change. This means that the accepted levels of defects, mistakes, or delays are no longer acceptable, and everyone involved is alerted to determine why those defects, mistakes, or delays exist and work together to rectify them.
- Stop depending on mass inspection and build quality into the product. In other words, the mass inspection neither improves nor guarantees quality.
- Stop awarding business on the basis of price and develop long-term relationships on performance.
- Constantly improve product, quality, and service. Any process or product variation is bad, and there are two types of variations: special variations and common variations. The special variations are within the control of the operator, and the common variations are beyond his/her control. About 94% of all variations fall under the category of common variations and are under the control of management [27]. Some of the causes of the common variations are hasty designs, inadequate testing of incoming materials, noise, humidity, and poor lighting.
- Institute training that includes modern methods. In developing a training program, factors such as objective identification, goal identification, employee identification, need analysis, and training of supervisors in statistical thinking should be considered.
- Employ modern supervision methods. Basically, this means teach and institute leadership.
- Eliminate fear. Past experience shows that many people find work unpleasant, not because they dislike doing their assigned task, but because of the environment under which they perform.
- Eliminate barriers between departments and emphasize team effort.
- Eradicate numerical goals, slogans, and posters that demand new productivity levels without enhancing the quality of approaches used.
- Remove numerical quotas and management by objectives (MBO).
- Eliminate barriers to employee pride in workmanship.
- Encourage a dynamic education and self-improvement program.
- Institute mechanisms to accomplish transformation.

8.5 TQM IMPLEMENTATION OBSTACLES AND GOALS FOR TQM PROCESS SUCCESS

The implementation of the TQM concept is not a straightforward task; an organization embarking on its implementation may face many obstacles as expressed in the form questions below [28]:

- Will senior management support introduction of TQM?
- How can the customer requirements be quantified?
- Who will set the TQM vision?
- How can it be possible to convince individuals of the need to change?
- How can it be possible to convince individuals that TQM is different?
- Is it possible to obtain support of engineers and their managers who possess an "independent" attitude?
- Does management clearly understand its purpose?
- Is there sufficient time available to implement TQM program?

The implementation of the TQM concept and its effectiveness are two different things altogether. In any case, for the effectiveness of the TQM process, some of the goals that must be fulfilled are the clear understanding of internal and external customer requirements by everyone in the organization, meeting of control guidelines per customer needs by all significant processes and systems, establishment of real rewards and incentives for employees when process control and customer satisfaction results are achieved, and employment of a system to constantly improve processes that better satisfy customers' current and future requirements.

8.6 QUALITY IN THE DESIGN PHASE

Even though the employment of the TQM concept should automatically help to improve product quality during the design phase, the performance of a specific task will certainly help to enhance quality of product design further. Thus, some of the important areas associated with improving the quality in design are described below [29].

8.6.1 PRODUCT DESIGN REVIEW

During the design phase of a product, there are several types of design review performed to improve the design from various aspects, and quality is one of them. These reviews include preliminary design review, detail design review, critical design review, preproduction design review, postproduction design review, and operations and support design review. Usually, there are several design reviews during product development, and the critical design review is performed to approve the final design.

The consideration of quality starts right from the preliminary design review and becomes stronger as the design progresses. The basic objective of quality assurance at the preliminary design review is to ensure that the design under consideration does not incorporate the problems of existing similar designs. This requires that the quality specialists involved must have a thorough knowledge of the strengths and weaknesses of the competing designs. Also, it must be remembered that, during the period of preliminary and the detail design reviews, about 95% of the product life cycle cost is established and the amount of money saved by preventing losses is at least 10 times the amount spent to perform these reviews [29].

Over the years, there have been various tools developed for use during the product design phase to improve quality. Such tools include quality function deployment (QDF), quality loss function, and benchmarking. These three tools are described subsequently in the chapter.

8.6.2 PROCESS DESIGN REVIEW

Soon after the approval of the preliminary design, the flowchart for the process is prepared. Thus, for some processes the analysis can start during the conceptual phase, and to assure the process works effectively at the desired level of yield, the professionals such as process, quality, and reliability engineers must work together as a team.

A method known as the failure mode and effect analysis (FMEA) is a useful tool to evaluate processes during the design phase. In order to achieve effective results, it is recommended that FMEA should be performed for the entire manufacturing process (i.e., including support services and operating procedures) rather than just for the equipment. Here, the basic objective is to make use of brainstorming to develop the best possible process. The performance of FMEA not only helps to identify process weaknesses, but also helps integration between quality assurance and manufacturing at the early stage. In any case, it may be said that FMEA helps to identify need and the effects of design change.

The FMEA procedure demands listing of potential failure modes of each part on paper and its effects on the listed subsystems. The steps involved in performing FMEA are listed below [30–32]:

- Construct a process flowchart that includes process inputs such as the materials and tooling. In addition, identify standard operating and maintenance procedures.
- List all process elements/components.
- Describe each component/element and identify its associated failure modes.
- Assign failure rate/occurrence probability to each component/element failure mode.
- List each failure mode cause and effect.

- Write remarks for each failure mode.
- Review each critical failure mode and take appropriate corrective measures.

It is to be noted that FMEA becomes failure modes, effects, and criticality analysis (FMECA) if criticalities or priorities are assigned to failure mode effects. In any case, the important characteristics of the FMEA are as listed below:

- An upward approach starts from the detailed level.
- It spots weak areas in design and highlights areas where additional or detailed analyses are necessary.
- By determining each component/element effects, the entire system is screened.
- It helps to improve communication among design interface personnel.

8.6.3 QUALITY ASSURANCE PLANS FOR ACQUISITION AND PROCESS CONTROL

The development of quality assurance plans for equipment acquisition and process control during the design phase helps to improve quality. The equipment procurement plan should include equipment performance verification with suppliers, testing for component interchangeability, statistical tolerance analysis, pilot runs, and so on. In turn, the plans for acquired components should be instituted prior to the start of pilot runs. In any case, in this regard, the important areas that warrant concern and cooperation are component qualification, closed-loop failure management, standard and screening tests, and implementation of process control throughout the production lines.

8.7 TQM TOOLS

There are a large number of tools and techniques available for practicing TQM in an organization. These include quality function deployment, quality loss function, Pareto diagram, cause-and-effect diagram, control charts, and benchmarking [26, 27]. All of these methods are described below [26, 27, 29].

8.7.1 QUALITY FUNCTION DEPLOYMENT (QFD)

QFD is a valued engineering tool used during product and process development and was developed by the Japanese in 1972 [27, 33]. In a broader context, QFD may be described as a formal process used for translating customer needs into appropriate technical requirements. In other words, QFD is a useful tool to assure that all operations of a firm are driven to satisfy customer needs rather than by senior management or by design specialists. The approach utilizes a set of matrices to relate consumer requirements to counterpart characteristics that are expressed as technical specifications and process control needs. The important QFD planning documents are a customer needs planning matrix, product characteristic deployment matrix, process plan and quality control charts, and operating instructions.

The purpose of the customer requirement planning matrix is to translate the consumer requirements into product counterpart characteristics, and on the other hand the product characteristic deployment matrix translates final product counterpart characteristics into crucial component characteristics. The function of the process plan and quality control charts is to identify important process and product parameters along with control points. The operating instructions are used to identify operations that must be accomplished to achieve critical parameters.

In a nutshell, the customer requirement planning matrix forms an important part of the QFD concept. Since the structure resembles a house, QFD is frequently called "The House of Quality." To build the house of quality, the following six steps are performed:

- Identify customer requirements.
- Identify the product/process characteristics that will satisfy the customer's wants.
- Relate the customer needs and counterpart characteristics.
- Perform an evaluation of competing products.
- Determine counterpart characteristics of competing products and establish goals.
- Choose counterpart characteristics to be used in the remaining process.

One important advantage of QFD is its usefulness in transferring customer needs (i.e., requirements plus product specifications) into either a technical product design or an organizational design. On the other hand, its major limitation is that the exact requirements must be highlighted in complete detail.

8.7.2 QUALITY LOSS FUNCTION

The quality loss function concept is attributed to Genichi Taguchi, a Japanese statistician and a Deming Prize winner. He argued that the quality cost goes up, not only because the finished product is outside given specifications, but also when it deviates from the set target value within the specifications. Thus, the basis for the quality loss function is that, if all parts are produced close to their given values, then it is reasonable to expect best product performance and lower cost to society. Another important factor in Taguchi's philosophy is that a product's ultimate quality and cost are determined, to a large degree, by its design and manufacturing processes. The quality loss function takes the general form of a simple quadratic formula and is expressed as follows:

$$L(w) = C(w - TV)^2 \tag{8.3}$$

where

w = the variable
$L(w)$ = the loss function at w
TV = the targeted value of the variable at which the manufactured product is expected to show its best performance
$(w - TV)^2$ = the square of the deviation from the target value
C = the proportionality constant

The above loss function is based on assumptions such as follows:

- There is a zero loss at the target value.
- The customer satisfaction is proportional only to the deviation from the target value.

The proportionality constant value may be determined by estimating the loss value for an acceptable deviation such as the tolerance limit. Thus, the following relationship may be utilized to estimate C:

$$C = \theta/\Delta^2 \tag{8.4}$$

where

Δ = the deviation from the target value, TV
θ = the loss expressed in dollars

8.7.3 PARETO DIAGRAM

The Pareto diagram is named after an Italian economist, Vilfredo Pareto (1848–1923). However, in the quality field Joseph M. Juran popularized Pareto's work by emphasizing that 80% of quality problems are the result of only 20% of the causes. Thus, the Pareto diagram is a useful tool to identify areas for a concerted effort.

Visually, the Pareto diagram is a type of frequency chart in which bars are arranged in descending order from left to right and provides order to activity. All in all, the Pareto principle could be a very useful tool in TQM effort and in particularly to improve quality of product designs, if applied correctly.

8.7.4 CAUSE-AND-EFFECT DIAGRAM

This diagram was originally developed by a Japanese quality expert, K. Ishikawa, in the early 1950s, thus also known as the Ishikawa diagram. In the published literature, this diagram is also referred to as a "Fishbone diagram" because its shape resembles the bones of a fish. The diagram is an extremely useful tool to perform cause-and-effect analysis to help in generating ideas and in finding the root cause of a problem for investigation. Visually, the right side of the diagram, i.e., the fish head, represents effect, and left of it, all possible causes are connected to the central "fish" spine.

In constructing the cause-and-effect diagram for the TQM effort, the effect is customer satisfaction and the major causes are materials, manpower, methods, and machines. These causes provide a good checklist for the initial stage analysis, and a properly developed diagram serves as an effective tool to highlight possible quality problems and inspection points. Some of the important steps involved in developing a cause-and-effect diagram are as follows:

- Establish problem statement.
- Brainstorm to identify possible causes.
- Develop major causes categories by stratifying into natural groupings and the steps of the process.
- Construct the diagram by linking the causes under appropriate process steps and write down the problem or the effect in the diagram box on the extreme right-hand side.
- Refine causes categories by asking questions such as what causes this and why does this condition exist?

Some of the benefits of the cause-and-effect diagram are useful to identify root cause, helpful to generate ideas, useful in guiding further inquiry, and present an orderly arrangement of theories.

8.7.5 CONTROL CHARTS

Control charts are used to highlight statistically significant changes that may happen in a process. They were first developed by Walter A. Shewhart of Bell Telephone Laboratories in 1924 to analyze discrete or continuous sets of data collected over a period of time [34].

These charts are graphic presentations of data collected over a time period, which show upper and lower limits for the process to be controlled. Furthermore, these charts allow a quick comparison to past performance data expressed in units of weight, length, temperature, pressure, and so on. All control charts are comprised of three lines: center line, upper control limit (UCL) line, and the lower control limit (LCL) line. The center line represents the process average and the data points outside UCL and LCL indicate that the process is out of control.

There are various reasons for using control charts with respect to improving product design quality, including to determine if a process is in statistical control, to provide a visual display of a process, to stop unnecessary process-related adjustments, and to provide information on trends over time. In developing a control chart, the factors such as listed below must be considered:

- characteristics to be examined
- the objective
- sample size and frequency
- the procedure to be used in selecting sample size and frequency

8.7.6 BENCHMARKING

Benchmarking may simply be described as a strategy of duplicating the best practice of a firm excelling in a specified business function. The history of the benchmarking may be traced back over a thousand years, when the Chinese General Sun Tzu [35] wrote, "Know your enemy and know yourself; in a hundred battles you will never be in peril." The basic idea in benchmarking is to develop a target or a goal at which to shoot and then to establish a benchmark against which to make comparison. Benchmarking may be divided into five different groups: competitive, industrial, shadow, internal, and world-class [27].

Competitive benchmarking is concerned with the identification of important competitive characteristics of a product or service and then comparing them to your own. Industrial benchmarking is also known as functional benchmarking. It is concerned with comparing functions within the same industrial sector; for example, some industrial associations exchange or publish information to make this comparison possible.

Shadow benchmarking is concerned with tracking a successful competitor's important product and service attributes and matching changes as they happen. This type of benchmarking is often practiced by gasoline stations and fast-food chains. Internal benchmarking is concerned with a firm examining its departments/divisions/branches and makes comparisons of repetitive operational functions. More specifically, those functions common to operating a current business are the basis for internal benchmarking.

The world-class benchmarking is concerned with comparing processes across diverse industries. In other words, the in-house process is compared to the best in the world, irrespective of industry. The benchmarks are developed through the following five steps [36, 37]:

- Identify the item to benchmark.
- Set up a team to carry out benchmarking.
- Highlight benchmarking partners.
- Obtain and perform analysis of benchmarking-related data.
- Take appropriate measures to meet or exceed the benchmark.

All in all, some of the common benchmark characteristics are difficulty in collecting data, speed and cost of performing benchmark, adoption risk, types of processes under investigation, difficulty in obtaining management backing to improvement ideas, and expected benefits.

0.0 TQM IMPLEMENTATION PITFALLS, SELECTIVE ORGANIZATIONS AND CENTERS, AND BALDRIGE AWARD WINNERS

Over the years, many companies have embarked on implementing TQM and have faced various difficulties. Some of the frequently occurring problems were as follows [2]:

- failure of senior management to devote sufficient time to the issue
- inadequate allocation of resources for developing and training manpower by the management
- insistence of management concerning process implementation in a manner the employees find unacceptable
- failure to delegate decision-making authority to lower organizational levels by the management

There are many organizations and centers concerned with promoting the TQM concept. Some of those are listed here [3]:

- American Society for Quality Control, 611 East Wisconsin Avenue, P.O. Box 3005, Milwaukee, Wisconsin 53201-3006
- American Society for Training and Development, 1640 King Street, Alexandria, Virginia 22313-9833
- Malcolm Baldrige National Quality Award, National Institute of Standards and Technology, A903 Administration Building, Gaithersburg, Maryland 20899

- American Productivity and Quality Center, 123 North Post Oak Lane, Houston, Texas 77024-7797
- Alliance of Organizational Systems Designers, 1365 Westgate Center Drive, Suite L-1, Winston-Salem, North Carolina 27103-2934
- Work in America Institute, 700 White Plains Road, Scarsdale, New York 10583
- Navy Personnel Research and Development Center, Quality Support Center, San Diego, California 92152-6800
- Quality and Productivity Management Association, 300 Martingale Road, Suite 230, Schamburg, Illinois 60173

Since the introduction of the Baldrige National Quality Award in 1987, many companies have been successful in receiving this prestigious recognition for their quality achievements. Those recipients [3] include Zerox Corporation, Westinghouse Electric Corporation (Commercial Nuclear Fuel Division), Motorola, Inc., IBM-Rochester, Federal Express Corporation, Cadillac Motor Car Division of the General Motors Corporation, Globe Metallurgical, Inc., Marlow Industries, Zytec Corporation, Wallce Company, Inc., Solectron Corporation, and Milliken & Company, Inc. It is interesting to note that, during the first years of this award, over 10,000 companies requested the application form, but only 66 returned the completed form for consideration. The award committee selected 13 firms for site visits and ultimately chose three as winners: Motorola, Inc., Commercial Nuclear Fuel Division of the Westinghouse Electric Corporation, and Globe Metallurgical, Inc.

8.9 SELECTED PUBLICATIONS ON TQM AND DESIGN QUALITY AND QUALITY MAGAZINES AND EDUCATIONAL VIDEOTAPES

Over the years hundreds of publications on TQM and design quality have appeared. A list of selected publications is as follows:

- Stein, R.E., *The Next Phase of Total Quality Management*, Marcel Dekker, Inc., New York, 1994.
- Tenner, R.R., Detoro, I.J., *Total Quality Management: Three Steps to Continuous Improvement*, Addison-Wesley, Reading, Mass., 1992.
- Gevirtz, C.D., *Developing New Products with TQM*, McGraw-Hill, New York, 1994.
- Crosby, P., *Let's Talk Quality: 96 Questions You Always Wanted to Ask*, Association for Quality and Participation, Cincinnati, Ohio, 1989.
- Mizuno, S., *Company-Wide Total Quality Control*, Asian Productivity Organization, Tokyo, 1989.

- Misuno, S., *Management for Quality Improvement: The Seven New QC Tools*, Productivity Press, Cambridge, Mass., 1988.
- Walton, M., *The Deming Management Method*, Association for Quality and Participation, Cincinnati, Ohio, 1986.
- Shores, A.R., *Survival of the Fittest: Total Quality Control and Management*, ASQC Quality Press, Milwaukee, Wisconsin, 1988.
- Burgess, J.A., Assuring the Quality of Design, *Machine Design*, February 1982, pp. 65–69.
- Chaparian, A.P., Teammates: Design and Quality Engineers, *Quality Progress*, Vol. 10, No. 4, 1977, pp. 16–17.
- Mann, N.R., *The Keys to Excellence: The Story of the Deming Philosophy*, Prestwick Books, Los Angeles, 1987.
- Ozeki, K., Asalea, T., *Handbook of Quality Tools: The Japanese Approach*, Productivity Press, Norwalk, Connecticut, 1990.
- Madu, C.N., Chu-hua, K., Strategic Total Quality Management, in *Management of New Technologies for Global Competitiveness*, Quorum Books, Westport, Connecticut, 1993, pp. 3–26.
- Schmidt, W.H., Finnigan, J.P., *The Race Without a Finish Line*, Jossey-Bass Publishers, San Francisco, 1992.
- Spenley, P., *World Class Performance, Through Total Quality*, Chapman and Hall, London, 1992.
- Ross, P.J., *Taguchi Techniques for Quality Engineering*, McGraw-Hill, New York, 1988.
- Lockner, R.H., Matar, J.E., *Designing for Quality*, ASQC Quality Press, Milwaukee, Wisconsin, 1990.
- Farquhar, C.R., Johnston, C.G., *Total Quality Management: A Competitive Imperative*, Report No. 60-90-E, 1990, Prepared by the Conference Board of Canada, Ottawa, Ontario, Canada.
- Evans, J.R., Lindsay, W.M., Quality and Product Design, in *Management and Control of Quality*, West Publishing Company, New York, 1989, pp. 188–221.
- Mears, P., TQM Contributors, in *Quality Improvement Tools and Techniques*, McGraw-Hill, Inc., New York, 1995, pp. 229–246.
- Joiner, B.L., The Statistician's Role in Quality Management, *Quality Progress*, January, 1988.
- Coate, L.E., TQM at Oregon State University, *Journal for Quality and Participation*, Dec. 1990, pp. 90–101.

From time to time, various journals and magazines publish material on TQM and design quality. In addition, many organizations have developed educational videotapes on the subject of quality. Some of those journals, magazines, and videotapes are as follows [3]:

- Quality Progress, Published by American Society for Quality Control, 310 West Wisconsin Avenue, Milwaukee, Wisconsin 53203
- The Quality Review, Published by American Society for Quality Control, 310 West Wisconsin Avenue, Milwaukee, Wisconsin 53203
- The Deming Library, Film, Inc., 5547 Ravenswood Avenue, Chicago, Illinois 60640-1199
- Juran on Quality Improvement, Juran Institute, 11 River Road, P.O. Box 811, Wilton, Connecticut 06897-0811
- The Letter, American Productivity and Quality Center, 123 North Post Oak Lane, Houston, Texas 77024
- The Quality Man with Philip Crosby, The Association for Quality and Participation, 801-B West 8th Street, Cincinnati, Ohio 45203-1601

8.10 PROBLEMS

(1) Compare the TQM concept with the traditional quality assurance program.

(2) Discuss at least five important elements of TQM.

(3) Describe the Deming approach to TQM.

(4) Describe in detail the FMEA approach.

(5) Discuss the following TQM methods:
- quality function deployment
- Pareto diagram
- cause-and-effect diagram
- benchmarking

(6) Discuss the following words with respect to TQM:
- total
- quality
- management

(7) Write an essay on the historical background of TQM.

8.11 REFERENCES

1. Sullivan, L.P., The Seven Stages in Company Wide Quality Control, *Quality Progress*, May 1986, p. 78.
2. Gevirtz, C.D., *Developing New Products with TQM*, McGraw-Hill, Inc., New York, 1994.
3. Schmidt, W.H., Finnigan, J.P., *The Race without a Finish Line: America's Quest for Total Quality*, Jossey-Bass Publishers, San Francisco, 1992.

4. Dobyns, L., Crawford-Mason, C., *Quality or Else*, Houghton Mifflin, Boston, 1991.
5. Walton, M., *Deming Management at Work*, Putnam, New York, 1990.
6. Van Ham, K., Setting a Total Quality Management Strategy, In *Global Perspectives on Total Quality*, The Conference Board, New York, 1991, p. 15.
7. Madu, C.N., Chu-hua, K., Strategic Total Quality Management (STQM), in *Management of New Technologies for Global Competitiveness*, edited by Madu, C.N, Quorum Books, Westport, Connecticut, 1993, pp. 3–25.
8. Spenley, P., *World Class Performance Through Total Quality*, Chapman and Hall, London, 1992.
9. Deming, W.E., *Out of Crisis*, MIT Center for Advanced Engineering Study, Cambridge, Massachusetts, 1986.
10. Farquhar, C.R., Johnston, C.G., *Total Quality Management: A Competitive Imperative*, Report No. 60-90-E, 1990. Available from the Conference Board of Canada, 255 Smyth Road, Ottawa, Ontario K1H 8M7, Canada.
11. Burati, J.L., Matthews, M.F., Kalidindi, S.N., Quality Management Organizations and Techniques, *Journal of Construction Engineering and Management*, Vol. 118, March 1992, pp. 112–128.
12. Peters, T.J., *Thriving on Chaos*, Harper and Row, Publishers, Inc., New York, 1987.
13. Aubrey, C.A., Felkins, P.K., *Teamwork: Involving People in Quality and Productivity Improvement*, Quality Press, Milwaukee, Wisconsin, 1988.
14. Crocker, O.L., Charney, S., Leung-chiu, J.S., *Quality Circles*, Penguin Press, New York, 1984.
15. Imai, M., *Kaizen, The Key to Japan's Competitive Success*, Random House, Inc., New York, 1986.
16. Dumas, R.A., Organization Wide Quality: How to Avoid Common Pitfalls, *Quality Progress*, Vol. 22, No. 5, 1989, pp. 41–44.
17. Matthews, M.F., Burati, J.L., *Quality Management Organizations and Techniques*, Source Document 51, The Construction Industry Institute, Austin, Texas, 1989.
18. Crosby, P.B., *Quality Is Free*, McGraw-Hill, Inc., New York, 1979.
19. Juran, J.M., *Juran on Planning for Quality*, The Free Press, New York, 1988.
20. Ledbetter, W.B., *Measuring the Cost of Quality in Design and Construction*, Publication 10-2, The Construction Industry Institute, Austin, Texas, 1989.
21. Ishikawa, K., *Guide to Quality Control*, Asian Productivity Organization, Tokyo, 1982.
22. Kume, H., *Statistical Methods for Quality Improvement*, The Association for Overseas Technology Scholarship, Tokyo, 1985.
23. Perisco, J., Team Up for Quality Improvement, *Quality Progress*, Vol. 22, No. 1, 1989, pp. 33–37.
24. Daetz, D., The Effect of Product Design on Product Quality and Product Cost, *Quality Progress*, June 1987, pp. 63–67.
25. Evans, J.R., Lindsay, W.M., *Management and Control of Quality*, West Publishing Company, New York, 1982.
26. Heizer, J., Render, B., *Production and Operations Management*, Prentice-Hall, Upper Saddle River, New Jersey, 1995.
27. Mears, P., *Quality Improvement Tools and Techniques*, McGraw-Hill, Inc., New York, 1995.

28. Klein, R.A., Achieve Total Quality Management, *Chemical Engineering Progress*, November 1991, pp. 83–86.
29. Raheja, D.G., *Assurance Technologies*, McGraw-Hill, Inc., New York, 1991.
30. Dhillon, B.S., Failure Modes and Effects Analysis: Bibliography, *Microelectronics and Reliability*, Vol. 32, 1992, pp. 719–732.
31. Dhillon, B.S., Singh, C., *Engineering Reliability: New Techniques and Applications*, John Wiley and Sons, New York, 1981.
32. Dhillon, B.S., *Systems Reliability, Maintainability and Management*, Petrocelli Books, Inc., New York, 1983.
33. Yoji, K., Editor, *Quality Function Deployment*, Productivity Press, Cambridge, Massachusetts, 1990.
34. *Statistical Quality Control Handbook*, Published by AT&T. Technologies, Indianapolis, 1956.
35. Tzu, S., *The Art of War*, Translated by S.B. Griffin, Oxford University Press, New York, 1963.
36. Spendolini, M.J., *The Benchmarking Book*, AMACOM, New York, 1992.
37. Camp, R., *Benchmarking*, ASQC Quality Press, Milwaukee, Wisconsin, 1989.

CHAPTER 9

Value Engineering and Configuration Management

9.1 INTRODUCTION

VALUE engineering may simply be described as a management technique used to perform analysis of an item function with the objective of achieving the desirable function at the minimum cost [1].

The history of value engineering goes back to the late 1940s. In fact, after World War II, some General Electric (GE) Company executives recognized in retrospect that various substitution of materials used because of wartime shortages of strategic materials resulted in products that were as good as the originals but less costly [2]. Consequently, in order to take advantage of these findings, they recommended the development of methods or techniques that would produce the same results deliberately. Thus, in 1947, Lawrence D. Miles, a GE electrical engineer, was selected to devise methods or techniques that—through material substitutions, changes in manufacturing approaches, or design—led to substantial savings. As a result of Miles's effort, the concept known as "Value Analysis" was born; in fact, it was in 1947 when the term *Value Analysis* was coined at the Lynn plant of GE. The application of this new concept saved millions of dollars for GE, and within 12 years of its applications, the number of full-time value analysts employed by the company mushroomed to 120 [3].

In 1952, the first "value analysis seminar" was conducted by GE, and in 1954 the Navy Bureau of Ships recognized the usefulness of value analysis and Admiral Wilson D. Leggett coined an alternative term *value engineering* [4]. In 1959, the Electronics Industries Association (EIA) organized the first national conference on value engineering, which was attended by over 250 people. In October of the same year, the Society of American Value Engineers (SAVE) was incorporated in Washington, D.C. Three years later, in 1962, SAVE started publishing a magazine entitled *The SAVE Journal of Value Engineering*. Today,

SAVE has thousands of members with a large number of chapters throughout the United States.

In 1983, the Japanese established the Miles Award for the most successful companies in reducing costs without sacrificing quality, and in 1985, the Japanese government awarded Lawrence D. Miles a High Order of Imperial Merit Medal for his contributions to value engineering/value analysis methods. Only three other American have received this medal: Edwards Deming, Lillian Gilbreth, and Peter Drucker [4]. In Japan presently, there are over 9000 practicing value analysts and a dynamic Society of Japanese Value Engineers.

In 1961, Lawrence D. Miles published a book [5] entitled *Techniques of Value Analysis and Engineering*. Today, there are many texts [1–12] available on the subject and hundreds of articles that contributed to the advancement of the value engineering discipline.

Configuration management may simply be expressed as a discipline for providing an organized mechanism to identifying, planning, controlling, and accounting for the status of configuration of an item, from its inception to retirement [13]. In the 1950s, the people involved in the missile launch program realized that the self-imposed disciplines practiced by equipment manufacturers and others were quite shallow with respect to documentation uniformity of configuration control and accounting. In fact, they discovered that the prototype units in orbit had no adequate documentation of changes made, part number identification records, and so on [14].

In 1962, after extensive research and consultation, the United States Air Force (USAF) released a document entitled, "Configuration Management During the Development and Acquisition Phases," AFSCM 375-1 and revised it in 1964 [15]. In the same year, the National Aeronautics and Space Administration (NASA) also published a similar document resembling AFSCM 375-1, entitled, *Apollo Configuration Management Manual*, NPC 500-1.

In 1965 and in 1967, the Army and Navy released their own documents entitled, *Configuration Management*, AMCR 11-26 and NAVMATINST 4130.1, respectively. This scenario was short-lived because in the late 1960s, Robert B. MacNamara, Secretary of Defense, issued a directive to have a single applicable configuration management procedure for all services [13]. Consequently, in 1968, the Department of Defense issued a document entitled, *Configuration Management*, DOD Directive 5010.8.

Some of the subsequent documents prepared by the Department of Defense were as follows [16]:

- MIL-STD-483, 1970, *Configuration Management Practices for Systems, Equipment, Munitions and Computer Programs*
- MIL-STD-482, 1970, *Configuration Status Accounting Data Elements and Related Features*

- MIL-STD-481 A, 1972, *Configuration Control-Engineering Changes, Deviations and Waivers* (short form)
- MIL-STD-1456, 1972, *Contractor Configuration Management Plans*

In 1992, the Department of Defense released another document entitled, *Configuration Management*, MIL-STD-973, and this document supersedes military standards 480-483, 1456, and 1521 [13].

Over the years, a large number of publications on configuration management has appeared. Some of those are listed in References [13–24]. However, it should be noted that today, the emphasis of publications is on software configuration management [25–38], rather than on hardware configuration management.

This chapter describes value engineering and configuration management separately.

9.2 VALUE ENGINEERING

This is a proven approach that should be considered in the development of every item. The application of the concept could result in 5–30% savings of the original total project cost [39]. Before we discuss value engineering any further, let us examine the meanings of the word *value*. Because this word may mean different things to different people, its first investigation of the meaning could be traced back to Aristotle, the teacher of Alexander the Great [8]. Around 350 B.C. Aristotle outlined seven classes of value: economic, moral, social, aesthetic, judicial, religious, and political. For our use, the economic classification is applicable. In any case, the *Oxford Dictionary* defines value as worth, desirability, or utility [40].

Miles used a deductive process to develop value analysis [4]. First, he developed a list of reasons for the unwarranted costs in products and processes, including lack of information and time, fear of personal loss, habits and attitudes, bias, other responsibilities of engineers, inclination to conform, lack of ideas and experience, preconceived ideas, and failure to make use of available specialists properly. On the basis of such factors, Miles and his team developed a philosophy with elements: the four classifications of value, the supplier workshop (an established purchasing approach), application of good human relations, and the value analysis job plan based on the classic scientific method of Anaxagoras [4].

Second, Miles established a new concept of evaluation by comparison and by stating that the value of anything is variable, depending on attitude and environment. Furthermore, he reasoned that value may be quantified at any time, only by comparing it with another item carrying out the same function.

The value engineering systematic approach often makes use of three basic terms: value, function, and product. Thus, prior to the detailed discussion of the value engineering concept, a clear understanding of these three terms is essential. The meaning of the word *value* was already described. Just like in the case of the word *value*, the term *function* means different things to different people, for example, duty, role, faculty, and office. However, from the value engineering perspective, we define it as "that which makes an item work or sell."

The word *product* also has a multiplicity of meanings [8]: the result of a chemical change, output of a farm, the result of a mathematical computation, and so on. From the perspective of value engineering, the word *product* is expressed as anything produced or obtained through the result of some operation or work, as by growth, generation, skill, study, or labor.

9.2.1 VALUE ENGINEERING PHASES

In implementing the value engineering concept, an organized and multidisciplined approach should be used. That approach is made up of seven distinct phases [39]:

- the team selection phase
- the information collection phase
- the brainstorming phase
- the alternative evaluation phase
- the alternative development phase
- the recommendation phase
- the implementation phase

Each of the above seven phases is discussed below separately.

The Team Selection Phase

A careful consideration in the selection of a value engineering team is essential; otherwise, the inclusion of wrong personnel can be absolutely fatal to the value engineering effort. One must also remember that, even though technical capability and experience are necessary qualifications, these alone may not make these personnel suitable for the engineering task.

Usually, a value engineering team is led by a leader, and the team members in turn are trained in the value engineering study process; thus, they possess the expertise to develop value engineering ideas. It is the team leader who normally decides the team size and the type of individuals needed for a specific value engineering project. In selecting team members, the past experience indicates that careful attention must be given to factors such as the representation of a cross section of technical fields, project cost management, and the composition of generalists and specialists, especially with respect to freethinking and flexibility. As far as the size of the team is concerned, according to Reference [4], five-member teams are the most effective.

The Information Collection Phase

This phase is concerned with collecting technical and cost-related information. This task could be divided into two categories: the information to be collected by the team members and the team leader. Some of the sources of collecting information by the team members are as follows:

- clients
- vendors
- similar projects

On the other hand, it is the team leader who collects information from each team member and then presents the analyzed information to the entire value engineering team.

The Brainstorming Phase

In this phase, on the basis of information collected, there are various cost-saving alternatives reviewed and discussed. In fact, the alternatives from the technical and cost angle are compared to the original design or function. Obviously, the creativity factor plays a crucial role in determining the desirable alternatives. The following four factors concerning individual value engineering team member characteristics are quite critical [3]:

- flexibility: This is concerned with an individual's ability to develop a wide variety of approaches or procedures to a problem in question.
- sensitivity: This is concerned with an individual's ability to recognize the existing problem.
- originality: This is concerned with an individual's ability to develop original solutions to the problem in question.
- fluency: This is concerned with an individual's ability to develop a substantial amount of alternative solutions to the problem in question.

On the other hand, the following three factors inhibit an individual's creative thinking [3]:

- emotional: This includes failure fear, desire for security, and so on.
- perceptual: This includes factors such as failure to use all the senses in observing the obvious.
- cultural: This includes pressures to conform to an existing pattern.

In order to encourage creativity within the value engineering group, the existence of an appropriate environment is necessary. The following actions on the part of the team leader will certainly help to generate a creative atmosphere [3]:

- Lead through persuasion rather than through orders.
- Encourage constructive nonconformity, individuality, and the like.
- Recognize and reward team member's accomplishments effectively.
- Encourage exchange of ideas and information among the team members and others involved on a regular basis.
- Continuously encourage and develop those with special talents.
- Allow team members to participate in decision making and in long-range planning.
- Keep the organization flexible enough to handle potential problems effectively.
- Keep updating the knowledge of involved individuals through courses, seminars, and so on.

There are many brainstorming methods that have been developed over the years. Probably the most widely used method is the group brainstorming. In a modern context the method was applied by Alex Osborn [41] in 1938; however, Hindu religious leaders have practiced it for over 400 years. These people called it Prai-Barshana: Prai means "outside yourself" Barshana means "question."

In this method a class of people participate in each brainstorming session. The length of the session may vary from 10 to 60 minutes—20 to 30 minutes is a good average. Usually, the best results are obtained when 8–12 persons participate in a brainstorming session. The method works because, during the session, one idea for solving a problem triggers another idea and the process continues. The following useful guidelines are associated with this method [42]:

- Members of the brainstorming group should belong to different backgrounds but similar interests.
- Strive for quantity. It means generate as many ideas as possible.
- Freewheel. It means the wilder the idea is, the better it is.
- Allow no criticism whatsoever during the session.
- Record ideas during the session.
- Keep the ranks of participants fairly equal.
- Select the session leader carefully.
- Appoint board (if applicable) to combine and improve ideas.

The Alternative Evaluation Phase

During this phase evaluations are performed to choose alterntives by keeping in mind two important factors: high degree of cost savings and ability to be implemented in the project's design. Once the alternatives are selected, the next step is to rank them according to their savings and acceptability to the client or customer (if applicable).

The Alternative Development Phase

In this phase each selected alternative is studied and analyzed to determine if the forecasted savings can be realized. Furthermore, for each of these alternatives, life cycle cost (LCC) analysis, including capital, operating, and maintenance cost studies, are performed.

The life cycle cost is defined as the sum of all costs incurred during the life span of an item, that is, procurement cost plus ownership cost. There are many different approaches available in published literature to estimate life cycle cost of an item [43]. In any case, the term *life cycle costing* was coined in 1965 [44].

The Recommendation Phase

In this phase, the team leader presents the fully developed alternatives to the team members. In turn, the team selects one or more alternatives as a recommendation to the client or to the others. Then, the recommendation is presented in a form of a report that includes information on various areas [4]: study objective, the team members' names and descriptions of their specialty areas, questionnaire data, function diagram accompanying costs and customer or user attitude allocations, value analysis target list, summary of proposals, and more.

The Implementation Phase

The final report of the recommendation phase plays a critical role during this phase. In fact, on the basis of this report, the recommended changes are implemented.

9.2.2 VALUE ENGINEERING POINTERS AND PROJECT SELECTION

Over the years people involved in value engineering work have developed many pointers/guidelines that help to improve the effectiveness of the value engineering concept application. Some of those pointers/guidelines are as follows [3]:

- Perform value analysis on specifics, rather than on generalities. It means select a specific part or area for value analysis studies and then focus effort intensely until a workable proposal or an alternative mechanism is developed.

- Obtain needed information from the best possible sources. It means the data to be used in value engineering studies must come from the most reliable and accurate sources, in order to develop the most effective value-related recommendations. The application of two basic principles are quite useful in this regard: obtain information from multiple sources and use the best possible source for the information desired.
- Shoot for team effort. Remember that, even though the value engineering studies require concerted individual effort, the end results could be enlarged many-fold with teamwork.
- Apply good business judgement. This means that, in order to make good value analysis–related decisions, the individual involved should make use of a philosophy such as, "If I would not spend my own money for it, it is probably too expensive" and "If it does not look right, it probably is not."
- Develop good human relations. The existence of good human relations in value engineering projects is quite crucial because of a high degree of dependence by the team members on each other. In fact, in this case good or bad human relations can mean success or failure of the project.
- Circumvent roadblocks. As the value analysis project progresses, the team usually forces various types of roadblocks. Some of those could be real and the others imaginary. The following two guidelines are useful to overcome such roadblocks: (1) Question each roadblock's validity. For example, ask questions such as, "Is it based on fact or opinion?" (2) Perform analysis of roadblocks to determine what is required to overcome them.

Furthermore, the value analysis team should also keep in mind that poor value develops in many ways [1]: lack of cost-reducing new ideas, lack of availability of essential information, nontroublesome items, personal inertia, failure to seek advice from others, temporary decisions, and predetermined reactions.

Because the items for a value engineering study may vary from a complete system to a software item, careful consideration in the selection process is required. In this regard, past experience has shown that the two general guidelines are often useful: (1) act where costs are high, and (2) act where profit is low.

In addition, in applying the value engineering concept to new products, factors such as listed below should be taken into consideration [3]:

- competition intensity
- value engineering requirement by the customer
- changes in feasibility
- value engineering performance measurement by the customer
- similarity of the new product to an old design
- required production units

- significance of cost to the customer
- future business prospects
- existing technical and performance problems
- design time tightness

All in all, during the selection of specific value analyses projects, there are many factors that should be carefully examined: relevant design history, the cost, the manufacturability, feasibility of sufficient return on investment, compatibility of the item cost with its function, production or procurement problems, and unreasonability of cost elements.

9.2.3 VALUE ENGINEER'S CHARACTERISTICS AND FUNCTIONS AND TEST FOR VALUE QUESTIONS

The job of the value engineer is very challenging as it relates to technical and management tasks [45]. Basically, a value engineer is concerned with investigation, fact collation, and ultimately, practical application after installation. In addition, some of the characteristics a value engineer should possess are substantial relevant technical experience, experience in cost analysis, good business judgement, being a team player, the ability to collect and organize information to identify potentially lucrative projects, flexibility and tactfulness, self-confidence, good emotional control, the ability to recognize different points of view, skill in both oral and written communication, and skill in other relevant human relation areas.

As it may be quite apparent from the required characteristics of a value engineer, this professional performs many different types of tasks: cost identification and reduction, risk identification, preparation of program plans and reports, motivation and indoctrination, reviewing part procurement, reviewing item simplification or elimination, liaison with other groups, and so on.

The pioneering value engineers prepared a list of questions to test for value that could also be applicable today to various items [3]:

- Does it require all its existing features?
- Is it possible to produce a usable part by a lower cost method?
- Does its application contribute to value?
- Is its cost proportional to its expected usefulness?
- Will anyone buy it for less?
- Is it possible to find a standards product/item/part that will be equally usable?
- Is it feasible that another reliable supplier can provide it cheaper?
- Is the item produced on correct tooling, considering its quantity?

- Is there anything more appropriate for the anticipated application?
- Do expected profit, material and labor costs, and overheads sum up to its cost, or is cost far greater than that?

9.3 CONFIGURATION MANAGEMENT

Nowadays, the term *configuration management* is fairly well known in the industrial sector, especially among those companies having contracts with the Department of Defense. Nevertheless, before we discuss configuration management any further, let us first examine the word *Configuration.*

Configuration may be expressed as the technical description and arrangement of items that are capable of fulfilling the fit, form, or functional needs outlined by the concerned product specification and drawings. In other words, configuration is simply a documentation that describes a new product design or, in fact, the technical detail of the arrangement of items that go into the makeup of that new design. Consequently, Reference [13] states that configuration management has the same meaning as *engineering documentation control.*

9.3.1 CHANGES TO PRODUCT AND IMPORTANT REASONS FOR A CONFIGURATION MANAGEMENT SYSTEM

Usually, during the life cycle of a product, various changes are made to it. One of the objectives of practicing configuration management is to ensure that such changes are properly documented. Those changes may be grouped into three categories [16]:

- changes that result in a significant savings in product life cycle cost
- changes that are absolutely desirable to eradicate product deficiencies
- changes that are useful in improving the product's or item's logistic support or its operational use

Specifically, in the product's or item's design and production phases, there could be various reasons for the changes [16]:

- Analysis or testing indicates deficiencies in the proposed design.
- Product or item designer has discovered a better way to meet the same objective.
- Production personnel find it difficult to produce the product/item on time according to the existing design.
- Better parts/items are available in the market.

- The production department has substituted parts, materials, or otherwise to reduce cost.
- The production department has discovered a mechanism to improve product performance; however, usually this is quite rare.

Obviously, these factors lead to the practicing of the configuration management concept. Thus, some of the specific reasons for having a configuration management system are as follows [13]:

- a good approach to control documentation
- technical complexity of modern products
- constant change in the state-of-the-art technology
- required by the new ISO 9000 series of standards on quality systems
- often, modern engineering products composed of both hardware and software
- complexity of customer and manufacturer interrelationships because of the technical complexity of the modern engineering products
- product liability suits
- the approach consisting of all the desirable elements for a good engineering documentation control mechanism

9.3.2 CONFIGURATION MANAGEMENT EFFORT GOALS AND SYMPTOMS OF INEFFECTIVE CONFIGURATION MANAGEMENT PRACTICE

The basic overall objective of practicing configuration management is to assure customers that the product under consideration is as originally expected, that is, functionally and physically, as defined by specifications and drawings. In order to satisfy this objective effectively, the configuration management effort's principal goals are as follows [15]:

- Establish definition of total documentation needed for product design, manufacture, and testing.
- Complete correctly the approved configuration descriptions, including specifications, operating procedures, lists of parts, test procedures, and applicable drawings.
- Complete accurately the identification of each part, subassembly, assembly, and material associated with the product.
- Keep accurate and complete preevaluation control and accounting of all the changes associated with product descriptions, as well as to the product itself.

- Establish a traceability mechanism for the end product and for its associated parts to their corresponding descriptions.

Basically, any firm involved in equipment manufacture practices some form of configuration management, especially when engineering revisions are involved. In such cases, the configuration management may or may not be effectively applied. Some of the principal symptoms of ineffective application of configuration management disciplines and procedures include unclear stated objectives, delayed decisions (irrespective of whether the decision is yes or no) on changes, inconsistent documentation with the equipment involved, and excessive cost of changes [46].

9.3.3 CONFIGURATION MANAGEMENT PLAN PRODUCT LIFE CYCLE PHASES

Under the configuration management plan, a product's life cycle is divided into four distinct phases as shown in Figure 9.1: concept formulation, definition, acquisition (this is further divided into two distinct stages, i.e., design and development, and production), and operational. As shown in the figure the baselines terminate such phases, that is, characteristics baseline the concept formulation phase, functional baseline the definition phase, and operational

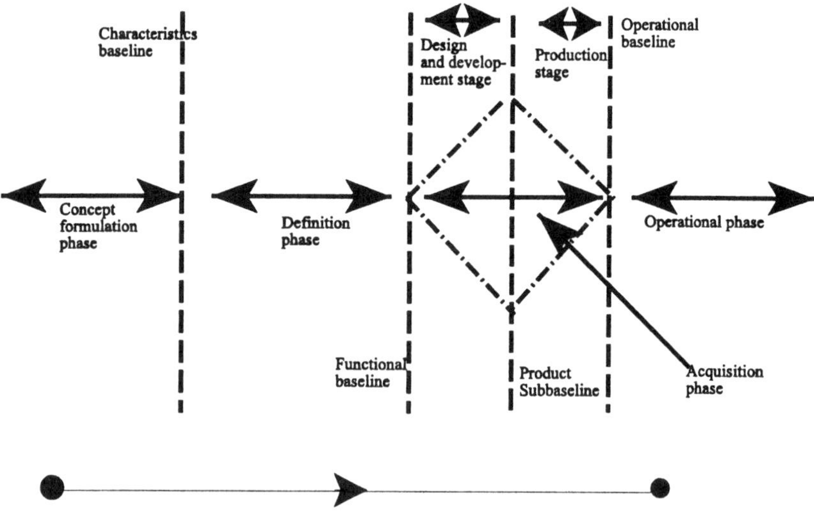

Figure 9.1 Product life cycle phases and baselines associated with the configuration management plan.

baseline the acquisition phase. In addition, the product subbaseline terminates the acquisition or procurement phase's design and development stage.

Broadly speaking, all the baselines denote checkpoints during the product life cycle. It means that the new phase cannot be started until all the details and questions raised during the previous phase are resolved satisfactorily. In addition to life cycle phases and baselines, the configuration management disciplines of identification, control, and accounting are applied. Each of these is described below separately.

Configuration Identification

Basically, this discipline requires that, during each life cycle phase, the complete and accurate description concerning the product must be documented. The important piece of documentation required under this discipline is the product specification. The other documentation includes items such as drawings, the bill of material, assembly part lists, spare part lists, manuals, and software documentation. Specifically, for example, the definition phase documentation items would be the specification, contract, schedule, and so on.

Configuration Control

This discipline is basically concerned with maintaining an effective control of work performed during a particular phase so that it is approved by the appropriate authorized bodies or individuals, as well as ensuring that such work is completely and accurately reflected in the baseline documentation. Furthermore, it may be said that the configuration control is an ongoing process that starts during the initial stages of a project and spills over to full service life of the item under consideration. Also, the configuration control is made up of those procedures through which changes are proposed, reviewed, coordinated, and either approved for incorporation or discarded altogether [15].

The configuration control effort prior to the acceptance of product by the customer is focused on control of configuration as outlined in documents such as specifications. After the first acceptance of the product by the customer, the configuration control effort is directed primarily toward hardware, with changes to documentation occurring because of approved specification and alterations made to hardware.

Configuration Status Accounting

This discipline is concerned with establishing records that allow the development of appropriate logistics support. These records include information on items such as follows:

- product location
- selected product items' identification by serial number makeup
- current modification status

Also, another function of the accounting discipline is auditing, which is implemented to ensure that the data describing the product is complete and accurate.

9.3.4 CONFIGURATION MANAGEMENT ORGANIZATION AND CONFIGURATION CONTROL BOARD

The configuration management organization is responsible for handling and controlling all engineering product changes in a company. However, the title of this organization may vary from one firm to another [13]: engineering change administration, configuration control, configuration management, engineering document control, and more. Similarly, the title of the person managing such an organization also varies: engineering change administrator, documentation manager, configuration manager, configuration control manager, and so on.

One of the most difficult tasks for senior management is to decide the location of the configuration management organization within the organizational setup. This is usually decided from meeting the needs of the organization in the best possible way. This is probably why, in some companies, the configuration organization reports to people such as the manager of quality, chief executive officer (CEO), the engineering services chief, and so on. However, in most firms, the configuration management organization is located within the engineering organizational structure [13].

In addition to the configuration management chief, other people who could be part of this organization are data entry clerk, drafter, secretary, technical writer, designer, and more. Obviously, these poeple support the configuration management chief, and their administrative responsibilities include [15]

- processing to completion all engineering change requests
- liaisoning with customers
- performing postrelease activities directed at the verification and approval of all engineering orders, specification revisions, and so on
- coordinating configuration control board (CCB) meetings and preparing meeting agendas and minutes

Configuration Control Board

Some organizations have found it useful to establish a configuration control board so that data and applicable arguments concerning proposed changes are presented by members representing all important functions. The board is the final authority within the project on all major proposed changes. It is important to note that the CCB is a nonvoting board, and the board chairman has the absolute authority to make decisions. However, in order for the board to become an effective group, it must meet regularly, for example, daily, twice a week, weekly, or monthly. Furthermore, the representations on the board must be from all the major functional areas within a firm for the best possible performance. The main function of a board member is to verify items such as

- the necessity of the change
- the feasibility of the implementation method
- the meeting of the schedule and cost requirement

The board member considers many factors with respect to the proposed change and its effects on other project areas, including configuration item's performance, impact on schedule and cost, configuration item's reliability, operator safety, configuration item's service life, part procurement-related problems, special manufacturing tool requirements, maintenance and repair problems, and configuration item's weight, size, and power consumption.

9.3.5 SUCCESSFUL CONFIGURATION MANAGEMENT FEATURES AND CONFIGURATION MANAGER'S FUNCTIONS AND QUALITIES

Just like in the case of any other discipline, there are specific characteristics or qualities that are also applicable to configuration management. One of the key features of the successful configuration management is meeting the goals stated earlier at a minimum cost to both manufacturer and customer. However, the important features that characterize successful configuration control, identification, and accounting are as follows [15]:

- complete and accurate configuration management goals, scope, and procedures descriptions
- efficient evaluation and processing of changes
- lowest level of labor needs
- complete and accurate change descriptions
- cooperative and responsive customers
- effective coordination among important components of the project team
- existence of a simple and straightforward configuration control, identification, and accounting procedure to obtain desirable results
- need to use a small number of forms and related documents to perform change implementation tasks, as well as to provide complete records of concerned changes

Just like any other engineering manager, the configuration manager has certain responsibilities, including configuration administration and planning CCB operations, data management, and people management. In order to perform tasks such as these effectively, the configuration manager must possess certain qualities: a good technical knowledge concerning configuration item design, testing, manufacturing, and quality control; skill in documentation; ability to administer and organize; diplomacy and tact; familiarity with the firm and the product; and attention to details. Furthermore, this person must be able to coordinate the work of individuals in widely different environments, capable of resisting pressures for shortcuts in the event of emergency, and capable of focusing on the entire picture of a project and at the same time not overlooking minute details.

9.3.6 SOFTWARE CONFIGURATION MANAGEMENT

There is probably more published literature available on software configuration management than on the hardware configuration management. In a few words software configuration may be described as the art of monitoring what has changed and how things are combined [47]. Furthermore, software configuration management does not differ significantly from general configuration mangement in terms of objectives, needs, and functions to be carried out. Another simple description of the software configuration management is provided by Reference [29]: software configuration management is simply general configuration management tailored to items that are largely made up of software.

The basic software configuration management requirements encompass four groups [47]: version control, process control, environment management, and build management. The version control is the process of monitoring changes to each file that supports parallel development by enabling straightforward branching and merging. In addition, each category or object that evolves under the software development environment ought to be version-controlled. The process control is a set of policies and enforcement procedures that ensure the development of the software per stated software development methodology. Also, the process control includes monitoring, notification, and access control. The environment management is the process that selects and presents each file's appropriate version to the developer in a manner that allows smooth functioning of the development tools. The build management is the process concerned with the development of software components, as well as producing a "bill-of-materials" that documents the contents of each software development.

Software Configuration Management Planning and Organization-Related Important Points

Some of these points are as follows [25]:

- The configuration management needs vary from project to project.
- The configuration management plan must effectively document every project's approach to configuration management, and in turn the

configuration management plan structure should be according to the Institute of Electrical and Electronic Engineers (IEEE) standard-82890.

- Manage two areas of configuration management very carefully: hand over of software to a maintenance team and the introduction of most appropriate configuration management practices to a wayward project.
- Ensure that the configuration management plan takes into consideration factors such as the use of the third party software, the project size, the item types to be managed, and the requirement for variants.
- Manage the configuration management plan as an item and refine it as the project progresses.
- Strictly apply two rules in evaluating configuration management tools: (1) the tool fully meets the project requirements, and (2) no configuration management tool is a panacea.

Software Configuration Management Common Pitfalls

Over the years, people working in software configuration management area have experienced various kinds of related hazards. Thus, common hazards such as listed below should be avoided to achieve effective configuration management application results [27]:

- excessive paperwork
- undocumented and circumvented procedures
- imcompatible documentation with programs
- software either overcontrolled or undercontrolled and controlled either too soon or too late
- existence of only one copy of a configuration-controlled master system
- premature acceptance of software developed by other people
- Performance of no low-level reviews prior to conducting high-level reviews
- delay in approving the development specification

9.4 PROBLEMS

(1) Write an essay on the historical developments of value engineering.
(2) Describe the seven phases of value engineering.
(3) What are the factors that should be taken into consideration in applying the value engineering concept to new products?
(4) Discuss the characteristics and functions of a typical value engineer.
(5) Discuss the historical developments of configuration management.

(6) What are the important reasons for having a configuration management system?
(7) List typical symptoms of practicing configuration management ineffectively.
(8) Discuss product life cycle phases and baselines associated with the configuration management plan.
(9) Describe the configuration control board.
(10) Discuss the features of practicing configuration management successfully.

9.5 REFERENCES

1. *Value Engineering, Engineering Design Handbook* AMCP 706-104, 1971, prepared by the United States Army Material Command, Washington, D.C.
2. Heller, E.D., *Value Management: Value Engineering and Cost Reduction*, Addison-Wesley Publishing Company, Reading, Massachusetts, 1971.
3. Brown, J., *Value Engineering*, Industrial Press, Inc., New York, 1992.
4. Fowler, T.C., *Value Analysis in Design*, Van Nostrand Reinhold Company, New York, 1990.
5. Miles, L.D., *Techniques of Value Analysis and Engineering*, McGraw-Hill Book Company, New York, 1961.
6. Oughton, F., *Value Analysis and Value Engineering*, Sir Isaac Pitman and Sons Ltd., London, 1969.
7. O'Brien, J.J., *Value Analysis in Design and Construction*, McGraw-Hill Book Company, New York, 1976.
8. Mudge, A.E., *Value Engineering*, McGraw-Hill Book Company, New York, 1971.
9. Fallon, C., *Value Analysis to Improve Productivity*, John Wiley & Sons, New York, 1971.
10. *Value Engineering*, Department of Defense Handbook 5010.8-H, September 1968, U.S. Government Printing Office, Washington, D.C.
11. Blyth, J.W., Woodward, R.G., *Application of Value Analysis/Engineering Skills*, Argyle Publishing Corporation, New York, 1967.
12. Ridge, W.J., *Value Analysis for Better Management*, American Management Association, New York, 1969.
13. Monahan, R.E., *Engineering Documentation Control Practices and Procedures*, Marcel Dekker, Inc., New York, 1995.
14. Hantz, E.G., Lager, A.E., Configuration Management: Its Role in the Aerospace Industry, *Proceedings of Product Assurance Conference*, American Society for Quality Control, Milwankee, WI, 1968, pp. 295–300.
15. Samaras, T.T., Czerwinski, F.L., *Fundamentals of Configuration Management*, John Wiley & Sons, New York, 1971.
16. Dhillon, B.S., *Systems Reliability, Maintainability and Management*, Petrocelli Books Inc., New York, 1983 (Chapter 12).
17. Laine, M.J., Spevak, E.C., Configuration Management, *Space/Aeronautics*, November 1966, pp. 74–81.

18. Feller, M., Configuration Management, *IEEE Transactions on Engineering Management*, Vol. 16, 1969, pp. 64–66.
19. Bunker, W.B., Objectives of Configuration Management, *Defense Industry Bulletin*, September 1967, pp. 1–3.
20. Seith, W., Configuration Management in Navy, *Defense Industry Bulletin*, April 1967, pp. 4–7.
21. Saynisch, M., Rugemer, D., *Configuration Management in Technological Projects*, Elsevier Science Publishers B.V., Amsterdam, 1986.
22. Samaras, T.T., *Configuration Management Deskbook*, Advanced Publication Consultant, Inc., New York, 1988.
23. Daniels, M.A., *Principles of Configuration Management*, Advanced Applications Consultants, Inc., San Francisco, 1985.
24. Eggerman, W.V., *Configuration Management Handbook*, Tab Books, Inc., Blue Ridge Summit, Pennsylvania, 1990.
25. Whitgift, D., *Methods and Tools for Software Configuration Management*, John Wiley & Sons, Chichester, 1991.
26. Babich, W.A., *Software Configuration Management*, Addison-Wesley Publishing Company, Reading, Massachusetts, 1986
27. Buckle, J.K., *Software Configuration Management*, The MacMillan Press Ltd., London, 1982.
28. Eggerman, W.V., *Configuration Management Handbook*, Tab Books, Inc., Blue Ridge Summit, Pennsylvania.
29. Berseff, E.H., Henderson, V.D., Siegel, S.G., *Software Configuration Management*, Prentice-Hall, Inc., Englewood Cliffs, New Jersey 1980.
30. Establier, J., Editor, *Software Configuration Management*, Springer-Verlag, Berlin, 1995.
31. Bersoff, E., Henderson, V., Siegel, S., Software Configuration Management: A Tutorial, *IEEE Computer Society Magazine*, Vol. 12, 1979, pp. 11–21.
32. Montes, J., Haque, T., A Configuration Management System and More! *Proceedings of the International Workshop on Software Version and Configuration Control*, Stuttgart, IEEE, New York, 1988, pp. 217–227.
33. Tichy, W., Tools for Software Configuration Management, *Proceedings of the International Workshop on Software Version and Configuration Control*, Stuttgart, IEEE, New York, 1988, pp. 175–196.
34. Wiebe, D., Generic Software Configuration Management: Theory and Design, Ph.D. Thesis, University of Washington, Seattle, 1990.
35. Buckley, F.J., *Implementing Configuration Management*, IEEE Press/IEEE Computer Society Press, New York, 1993.
36. Sprague, K., The Role of Software Configuration Management in a Measurement-Based Software Engineering Program, *ACM SIGSOFT Engineering Notes*, Vol. 16, 1991, pp. 62–66.
37. Ronald Berlack, H., *Software Configuration Management*, John Wiley and Sons, New York, 1992.
38. Tichy, W.F., Editor, *Configuration Management*, John Wiley and Sons, Chichester, 1994.
39. Archarya, P., Value Engineering, *Journal of Management in Engineering*, November/ December 1995, pp. 13–17.

40. Fowler, H.W., Fowler, F.G., *The Concise Oxford Dictionary of Current English*, Oxford University Press, London, 1964.
41. Osborn, A.F., *Applied Imagination*, Scribner and Sons, New York, 1963.
42. Dhillon, B.S., *Engineering Management*, Technomic Publishing Company, Lancaster, Pennsylvania, 1987.
43. Dhillon, B.S., *Life Cycle Costing: Techniques, Models and Applications*, Gordon and Breach Science Publisher, New York, 1989.
44. *Life Cycle Costing in Equipment Procurement*, Report No. LMI Task 4C-5, Prepared by Logistics Management Institute (LMI), Washington, D.C., April 1965.
45. Clawson, R.H., *Value Engineering for Management*, Auerbach Publishers, Inc., New York, 1970.
46. Hajek, V.G., *Management of Engineering Projects*, McGraw-Hill Book Company, New York, 1977, Chapter 13.
47. Leblang, D.B., Levine, P.H., Software Configuration Management: Why Is It Needed and What Should It Do? In *Software Configuration Management*, edited by J. Estublier, Springer-Verlag, Berlin, 1995, pp. 53–60.

CHAPTER 10

Life Cycle Costing

10.1 INTRODUCTION

TODAY life cycle costing is an important element during product design and development in many sectors of industry. The life cycle cost of a product is the total cost (i.e., procurement plus ownership) over its life span. The life cycle costing concept may have been practiced in one form or the other earlier, but it was just over 30 years ago when the term *life cycle costing* was first coined in a military-related document [1]. This document was prepared by the Logistics Management Institute (LMI), Washington, D.C. for the Assistant Secretary of Defense for Installations and Logistics [2]. As the result of this document, the Department of Defense published a number of guidelines in the early 1970s: *Life Cycle Costing Procurement Guide* (interim) [3], *Life Cycle Costing in Equipment Procurement—Casebook* [4], and *Life Cycle Costing Guide for System Acquisitions* (interim) [5].

In 1971, a directive [6] that instituted the requirement for life cycle costing in the procurement of major defense systems was issued by the Department of Defense. In 1974, the state of Florida formally adopted the application of life cycle costing, and the following year, the U.S. Department of Health, Education, and Welfare initiated a project entitled "Life Cycle Budgeting and Costing as an Aid in Decision Making." In 1978, the National Energy Conservation Policy Act, requiring every new federal building to be life cycle cost-effective, was passed by the U.S. Congress. Since the mid-1970s, many state governm-ents have passed legislation that requires the performance of life cycle cost analysis to be mandatory in the planning, design, and construction of state buildings (e.g., Texas, Alaska, New Mexico, Maryland).

In 1981, a survey article on life cycle costing was published [7], which listed most of the important publications on life cycle costing. In addition, over the years, a large number of publications on the subject have appeared [2, 8–20].

This chapter presents various aspects of the life cycle costing concept and related areas.

10.2 REASONS FOR SYSTEM COST INCREASES AND LIFE CYCLE COST ANALYSIS PURPOSES

Today, many systems and products have been experiencing a continuous increase in cost because of factors such as economic trends and rising inflation. This increase in cost is not only concerned with the product acquisition, but also with its operation and maintenance. Specifically, there are many causes for product cost growth [19]: engineering changes occurring throughout the design and development phase because of improving performance, adding capability, or the like; production/construction changes; changing suppliers in the acquisition of product parts; changes in the logistic support capability; changes in estimating procedures and inaccuracies in initial estimates; and unforeseen problems.

Also, there are many additional problems associated with the actual determination of product cost: unrealistic accounting procedures; inflexible budgeting practices; often invisible total system cost, in particular, the costs associated with system operation and support; and incorrect application of individual factors in cost estimation.

The source of the purpose of performing product life cycle cost analyses includes [19]

- evaluation of alternative operational, utilization, and environmental profiles
- evaluation of alternative acquisition source selection for a specified item
- evaluation of alternative logistics support plans
- evaluation of alternative production approaches
- evaluation of alternative product disposal and recycling procedures
- evaluation of alternative product maintenance concepts
- evaluation of alternative product distribution channels
- evaluation of alternative product design configurations

Additional purposes of life cycle costing include determination of cost drivers, forecasting future budget needs, formulating contractor incentives, optimizing training needs, and providing goals for program controls [21].

10.3 DATA REQUIRED FOR LIFE CYCLE COSTING AND DATA SOURCES

Life cycle costing studies require various kinds of data. Prior to starting a life cycle costing study, it is quite useful to seek answers to questions on various areas, including estimated goal, required data, assumptions, ground rules, constraints, study users, analyst's responsibility, required analysis details and format, estimating procedures, treatment of uncertainties, required precision and accuracy, time schedule, involved personnel, and fund limitations [9]. Specifically, information such as listed below is required to perform a life cycle costing study of an item [13]:

- acquisition cost
- disposal cost or sale income
- maintenance cost per annum
- discount and escalation rates
- delivery cost
- installation cost
- expected life in years
- operating cost per annum that includes material cost, cost of supplies, energy cost, cost of labor, and insurance
- taxes (e.g., investment tax credit, tax benefits from depreciation)

There are many different sources of data for life cycle costing: market analysis data, consumer utilization data, reliability and maintainability data, engineering design data, logistic support data, accounting data, production or construction data, value analysis and related data, and management planning data [8].

Also, there are many publicly available data sources from which information required for life cycle cost studies may be retrieved. However, care must be taken when using such information. Often, adjustment factors are applied to such data to compensate for any difference in anticipated operational environments, technology, configuration, time frame, and so on. This category of data includes standard cost factors that cover areas such as overhead rate ($ per direct labor cost), engineering labor cost ($ per labor hour), material cost ($ per pound or per foot), manufacturing labor cost ($ per labor hour), shipping cost ($ per pound per mile), fuel cost ($ per gallon), cost of facilities ($ per cubic foot occupied), training cost ($ per trainee week), and cost of maintaining inventory (percent of the inventory value per year). Prior to anticipating using any cost data in life cycle cost studies, answers to questions on various areas should be sought: data bias, data availability, data applicability, data obsolescence, data comparability to other existing data, and data orientation towards the problem in question [2].

In the event an organization contemplating to develop its own cost data bank first of all should seriously consider factors such as flexibility, responsiveness, orientation, ready accessibility, comprehensiveness, size, uniformity, and expansion or contraction capability. In addition, at minimum such a data bank should include information such as cost records, user pattern records, procedural records: operation and maintenance, and descriptive records: hardware and site.

10.4 LIFE CYCLE COST ANALYSIS STEPS AND ASSOCIATED ACTIVITIES, AND AREAS FOR LIFE CYCLE COSTING PROGRAM EVALUATION

Different people have presented different steps for performing life cycle cost analysis. A typical list of such steps follows [22]:

- Estimate the expected life of the product under consideration.
- Obtain estimates for all concerned costs, including operation and maintenance costs.
- Estimate product's terminal/disposal value/cost.
- Subtract/add the terminal/disposal value/cost from/to the product's ownership cost.
- Discount the resulting amount of the preceding step to present value.
- Add the acquisition cost to the amount obtained in the preceding step to estimate product life cycle cost.
- Repeat the above steps for every competing product being considered for procurement.
- Make comparisons among life cycle costs of products under consideration.
- Procure the product with the least life cycle cost.

There are many activities that are associated with life cycle cost analysis: identification of cost drivers, performance of sensitivity analyses, establishment of constant dollar cost profiles, determination of cause-and-effect relationships, establishment of escalated and discounted life cycle costs, establishment of definitions of activities that produce product ownership costs, development of cost estimating relationships for each component in the life cycle cost breakdown structure, and establishment of an accounting breakdown structure [10].

In order to assure the effectiveness of the life cycle costing program, it is reviewed periodically by the product manufacturer/user management. There are many areas in which questions could be raised to determine the life cycle costing program adequacy. Those areas include incorporation of life cycle cost-related requirements into design subcontracts; cost estimating data bank broadness; cost driver identification; performance of trade-off studies; satisfactory consideration of inflation and discounting factors; effectiveness of cost-estimating methods, reliability, maintainability, and system safety programs compatibility with life cycle cost requirements; planning and design group awareness of concerning life cycle cost targets; cost model construction; cost-performance reviews of subcontractors; validation of cost estimates; buyer awareness of top ten cost drivers; coordination of life cycle cost and design to cost efforts; design group's promptness concerning cost estimates; review of the top ten cost drivers for economy by management; and the qualifications of the life cycle costing management representative [9, 23].

10.5 COSTS GENERATED OVER THE PRODUCT LIFE CYCLE

There are various kinds of actions taken over the product life cycle. The implications of many such actions affect product life cycle cost, especially of those actions taken at the initial stages of the product life cycle. According to Reference [19], approximately 60% of the projected life cycle cost of a product is committed by the end of the planning and conceptual design stage, even though at this point in time, the actual amount spent on the project is relatively low. Usually, the life cycle cost falls into four categories based on organizational activity over the product life cycle: research and development cost, production and construction cost, operation and support cost, and retirement and disposal cost [8].

The research and development cost includes the cost of items such as product research, engineering design, software, feasibility studies, initial planning, design data and documentation, market analysis, and test and evaluation of engineering models. The production and construction cost is composed of the costs of items such as manufacturing (i.e., test, assembly, and fabrication), production operations, facility construction, operations analysis, and process development.

The elements of the operation and support cost are product distribution cost (i.e., cost of marketing and sales, transportation, etc.) and the cost of sustaining maintenance and logistic support throughout the product life cycle (this cost is associated with items such as maintenance, customer service, technical data, supply support, test equipment, and so on).

The retirement and disposal cost consists of elements such as material recycling cost, applicable logistic support requirements cost, and the cost associated with the disposal of nonrepairable items throughout the product life cycle. As is evident, during the product life cycle, various kinds of cost-related tasks are performed. Obviously, the main goal of these tasks is to reduce life cycle cost. The discussion on this subject is broken into four product life cycle categories:

(1) Conceptual design phase: Usually, during this phase, quantitative cost figures of merit are developed as requirements. These requirements serve as targets for the product to be designed, tested, produced, and supported. Also, usually, it is beneficial to adopt the design-to-cost (DTC) concept to establish cost as a product design constraint along with other constraints such as weight, performance, reliability, and so on.

(2) Preliminary design phase: After the establishment of quantitative cost requirements, the next step includes an iterative process of synthesis, optimization and trade-off, and the like. As the product development progresses, various competing options are considered in leading to a preferred product configuration. Life cycle cost analyses of each alternative are conducted by considering two objectives: (1) the compatibility of the selected candidates with the set cost goals and (2) preferred overall cost-effectiveness of candidates being considered. During this phase, life cycle cost analysis serves as a useful evaluation tool to conduct various trade-off studies.

(3) Detail design and development phase: In this phase, as product design is refined further, the more accurate design data become available. Thus, the life cycle cost analyses are concerned with the determination of specific design characteristics, the cost-generating variables' prediction and cost estimation, and the life cycle cost projection as a cost profile. The end results are compared to the set goals, and then corrective measures are taken as appropriate.

(4) Production, utilization, and support phase: In this phase, the cost concerns are addressed through data collection, analysis, and assessment function. Also, major cost drivers are identified along with cause-and-effect relationships, in addition to gaining and utilizing valuable information for the purposes of product improvement.

10.6 LIFE CYCLE COSTING PERFORMANCE SKILLS AND ASSOCIATED PROFESSIONALS

The performance of life cycle cost analysis is not an easy task. The individuals conducting such tasks require knowledge and skills in many areas, including finance and accounting, estimating, logistics, engineering, reliability and maintainability, quality control, statistical analysis, contracting, and manufacturing engineering. However, usually it is quite rare that an individual performing life cycle costing studies has skills in all such disciplines; thus, it takes assistance, as it becomes appropriate, from other concerned specialists. Usually, such specialists are design engineers, manufacturing engineers, logistics engineers, tooling engineers, planning engineers, test engineers, reliability and maintainability engineers, and quality control engineers.

10.7 LIFE CYCLE COST ESTIMATION MODELS

The basic objective of all life cycle cost estimation models is to estimate total cost of an item over its entire life span. Over the years, many different types of life cycle cost estimation models have been developed. Many such models are not only tailored for a specific product/equipment/system, but different models are used to estimate life cycle cost within some organization. For example, within the U.S. Department of Defense, Army and Navy use two different life cycle cost models.

The life cycle cost (LCC_N) expressed by the Navy model is as follows [10]:

$$LCC_N = RDC + IC + OSC + ASC + TC \tag{10.1}$$

where

RDC = the research and development cost
IC = the investment cost
OSC = the operating and support cost
ASC = the associated systems cost
TC = the termination cost

The life cycle cost (LCC_A) expressed by the Army model is as follows [10]:

$$LCC_A = RDC + IC + OSC \tag{10.2}$$

Reference [8] has expressed Equation (10.2) slightly differently:

$$LCC = RDC + IC + OMC \tag{10.3}$$

where

LCC = the life cycle cost
OMC = the operations and maintenance cost

The detailed breakdowns for each of the three terms in the right-hand side of Equation (10.2) are presented below.

10.7.1 RDC

The components of the research and development cost (RDC) are as follows:

- cost of advanced research and development (ARDC)

- cost of equipment development and tests (EDTC)
- cost of engineering design (EDC)
- cost of program management (PMC)
- cost of engineering data (EDAC)

The cost of advanced research and development (ARDC) is estimated from the following relationship:

$$\mathrm{ARDC} = \sum_{j=1}^{m} \mathrm{ARDC}_j \tag{10.4}$$

where

m = the number of activities
ARDC_j = the cost of specific activity j

The cost of equipment development and tests (EDTC) is expressed by

$$\mathrm{EDTC} = \sum_{j=1}^{k} \mathrm{C}_j + \mathrm{PFAC} + \mathrm{PMC} \tag{10.5}$$

where

k = the number of identifiable tests
C_j = the test operation cost and support associated with specific test j
PFAC = the prototype fabrication and assembly labor cost
PMC = the prototype material cost

The cost of engineering design is given by

$$\mathrm{EDC} = \sum_{j=1}^{x} \mathrm{EDC}_j \tag{10.6}$$

where

x = the number of design activities
EDC_j = the cost of specific design activity j

The cost of program management is expressed by

$$\mathrm{PMC} = \sum_{j=1}^{m} \mathrm{PMC}_j \tag{10.7}$$

where PMC_j = the cost of specific activity j

The cost of engineering data (EDAC) is given by

$$\text{EDAC} = \sum_{j=1}^{y} \text{EDAC}_j \tag{10.8}$$

where

y = the number of data items
EDAC_j = the cost of specific data

10.7.2 IC

The three major elements of the investment cost (IC) are as follows:

- cost of product/system construction (PCC)
- cost of product/system manufacturing (PMC)
- cost of initial logistic support (ILSC)

The major components of the product/system construction cost (PCC) are the costs associated with manufacturing facilities, test facilities, the acquisition of operational facilities, and maintenance facilities.

The two main elements of the cost of product/system manufacturing (PMC) are recurring and nonrecurring manufacturing costs. There are many components of the initial logistic support cost (ILSC): initial transportation and handling cost, provisioning cost, initial training and training equipment cost, logistic program management cost, initial inventory management cost, operational test and support equipment acquisition cost, initial spare/repair part material cost, and technical data preparation cost.

10.7.3 OMC

The four major components of the operations and maintenance cost are expressed as follows:

- item/equipment life cycle maintenance cost (ILMC)
- item/equipment life cycle operations cost (ILOC)
- item/equipment modifications cost (IMC)
- item/equipment phase-out and disposal cost (IPC)

The item/equipment life cycle maintenance cost (ILMC) is estimated from the following relationship:

$$\text{ILMC} = \sum_{i=1}^{6} \text{C}_i \tag{10.9}$$

where

C_1 = the technical data cost
C_2 = the maintenance personnel and support cost
C_3 = the maintenance facilities cost
C_4 = the test and support equipment maintenance cost
C_5 = the spare/repair parts cost
C_6 = the transportation and handling cost

The components of the item/equipment life cycle operations cost (ILOC) are as follows:

- operational facilities cost
- cost of operating manpower
- support and handling equipment cost
- operator training cost

The following relationship can be used to estimate item/equipment modifications cost (IMC):

$$\text{IMC} = \sum_{j=1}^{y} k_j \tag{10.10}$$

where

y = the number of item/equipment modifications
k_j = the cost associated with specific modification j

The item/equipment phase-out and disposal cost (IPC) can be estimated using the following formula:

$$\text{IPC} = nf(\text{IDC} - \text{RV}) \tag{10.11}$$

where

IDC = the item/equipment disposal cost
RV = the reclamation value
n = the number of corrective maintenance actions
f = the condemnation factor

10.7.4 SPECIFIC LIFE CYCLE COST ESTIMATION MODELS

Over the years, there have been many specific life cycle cost models developed for specific systems (e.g., switching power supply [24], electric motor [25], and early warning radar [2]). Many of such models may be found in Reference [2].

Life Cycle Cost of a Switching Power Supply

This is expressed by [24]

$$\mathrm{LCC_S} = \mathrm{IC} + \mathrm{FC} \tag{10.12}$$

where

$\mathrm{LCC_S}$ = the life cycle cost of switching power supplies
IC = the initial cost
FC = the failure cost

In turn, the failure cost (FC) is given by

$$\mathrm{FC} = K\lambda(\mathrm{RC} + \mathrm{SC}) \tag{10.13}$$

where

K = the product's expected life
λ = the product's estimated failure rate
RC = the repair cost
SC = the spare cost

The spare cost is expressed by

$$\mathrm{SC} = \mathrm{SC_u}\alpha \tag{10.14}$$

where

$\mathrm{SC_u}$ = the unit spare cost
α = the fractional number of spares for each active unit

Life Cycle Cost of an Electric Motor

Under the assumption of negligible maintenance cost, the life cycle cost of an electric motor is expressed by [25]

$$\mathrm{LCC_m} = \mathrm{PC} + \mathrm{OC} \tag{10.15}$$

where

$\mathrm{LCC_m}$ = the life cycle cost of an electric motor
PC = the motor procurement cost
OC = the motor operating cost

The present value, PV_i, of the motor operating cost, OC_i, for year i is expressed by

$$PV_i = OC_i(1 + j)^{-i} \tag{10.16}$$

where j is the compound interest rate fraction.

The total present value, TPV, of the motor operating cost for estimated m operational years is expressed by

$$\begin{aligned} TPV = {} & OC_1(1 + j)^{-1} + OC_2(1 + j)^{-2} + OC_3(1 + j)^{-3} \\ & + \cdots + OC_m(1 + j)^{-m} \end{aligned} \tag{10.17}$$

Assuming that the motor operating cost, OC_i, varies insignificantly during the motor useful operational life and that its effect is considered through an escalation factor, modifying Equation (10.17), we have

$$\begin{aligned} TPV = {} & OC[(1 + \theta)^1(1 + j)^{-1} + (1 + \theta)^2(1 + j)^{-2} + (1 + \theta)^3(1 + j)^{-3} \\ & + \cdots + (1 + \theta)^m(1 + j)^{-m}] \end{aligned} \tag{10.18}$$

It is to be noted that in Equation (10.18) we have assumed $OC = OC_1 = OC_2 = \cdots = OC_m$ and θ donates escalation rate fraction. The simplified version of Equation (10.18) is as follows:

$$TPV = (OC)\mu\left(\frac{1 - \mu^m}{1 - \mu}\right) \tag{10.19}$$

where

$$\mu = \left(\frac{1 + \theta}{1 + j}\right)$$

The following formula can be used to estimate yearly motor operating cost, OC, in dollars:

$$OC = (OH)S(EC)(0.746)/1000(ME) \tag{10.20}$$

where

ME = the motor efficiency fraction
S = the motor size expressed in horsepower (hp)
OH = the annual motor operating hours
EC = the electricity cost expressed in dollars per megawatt hour (MWh)

10.8 COST ESTIMATION MODELS

In order to estimate the cost of specific life cycle costing activities, over the years various types of cost estimation models have been developed. This section presents some of those models.

10.8.1 POWER LAW AND SIZING COST ESTIMATION MODEL

This is a useful model, in many cases, to estimate equipment cost. The equipment under consideration must be similar in type to reference equipment and varies only in size. Thus, the cost of the new equipment under design, C_{nd}, may be estimated from the following relationship [2, 8]:

$$C_{nd} = C_{rd}\left(\frac{S_{nd}}{S_{rd}}\right)^{n} \tag{10.21}$$

where

C_{rd} = the known cost of the referenced equipment of size S_{rd}
S_{rd} = the referenced equipment design size
S_{nd} = the design size of the new equipment under consideration
n = the cost capacity factor, $0 \leq n \leq 1$. Obviously, a linear relationship exists at $n = 1$; thus, the economies of scale do not apply and for much equipment the value of m is around 0.5. However, in the case of chemical processing equipment, the approximate value of m is 0.6

Example 10.1

Assume that the cost of a 20-horsepower (hp) reference motor is \$800. Estimate the cost of a new similar 25-hp motor under design by assuming the value of the cost capacity factor is 0.5.

By substituting the given data into Equation (10.21), we get

$$\begin{aligned} C_{nd} &= (800)\left(\frac{25}{20}\right)^{0.5} \\ &= \$894.427 \end{aligned}$$

Thus, the cost of the new motor under design will be approximately \$895.

10.8.2 CORRECTIVE MAINTENANCE LABOR COST ESTIMATION MODEL

The corrective maintenance labor cost is one of the most important elements in equipment life cycle costing. It is related to equipment mean time between failures (MTBF) and mean time to repair (MTTR). The corrective maintenance labor cost is expressed by [2]

$$C_{cm} = (\mathrm{SOH})\mathrm{LC_m}(\mathrm{MTTR})/\mathrm{MTBF} \tag{10.22}$$

where

C_{cm} = the yearly cost of the corrective maintenance
SOH = the equipment scheduled operating hours
$\mathrm{LC_m}$ = the corrective maintenance labor cost per hour

Example 10.2

An electric motor's MTBF and MTTR are 1500 hours and 6 hours, respectively. Annually, the motor is scheduled for 5000 hours of operation. If the cost of maintenance labor is $30 per hour, calculate the annual cost for performing corrective maintenance on that motor.

Substituting the specified data into Equation (10.22) yields:

$$C_{cm} = (5{,}000)(30)(6)/(1500)$$
$$= \$600$$

Thus, the annual cost of performing motor corrective maintenance will be $600.

10.8.3 SOFTWARE DEVELOPMENT COST ESTIMATION MODEL

Nowadays, many engineering systems use various kinds of computer software. In these systems, often the cost of software development becomes a significant element of the system life cycle cost. The total cost of developing software is expressed by [2, 26, 27]:

$$\text{TSDC} = \text{SDC}_\text{P} + \text{SDC}_\text{S} \tag{10.23}$$

where

TSDC = the total software development cost
SDC_P = the primary software development cost
SDC_S = the secondary software development cost

The primary software development cost (SDC_P) is expressed by

$$\text{SDC}_\text{P} = r(\text{MM}) \tag{10.24}$$

where

MM = the manpower required to develop software, expressed in man-months, including activities such as design, analysis, code, debug, test, and checkout.

r = the average labor rate of the manpower used to develop software, expressed in dollars per man-month, including costs such as general, administration, overhead, and appropriate fees.

The secondary software development cost (SDC_S) is given by

$$\text{SDC}_\text{S} = \sum_{i=1}^{m} \text{RDC}_{\text{S}i} \tag{10.25}$$

where

m = the number of secondary resources
RDC_{Si} = the cost associated with secondary resource i, expressed in dollars

10.8.4 PARALLEL SYSTEM OPERATION, MAINTENANCE, AND FAILURE COST ESTIMATION MODEL

This model is concerned with estimating the minimum operation, maintenance, and failure cost of a parallel system composed of n units [28]. The total parallel system cost due to operation, maintenance, and failures is expressed by

$$TC_P = \frac{(OMC)(UT)n + (DTC)DT}{UT + DT} \quad (10.26)$$

where

TC_P = the total parallel system operation, maintenance, and failure cost (per unit time)
UT = the parallel system uptime
DT = the parallel system downtime
DTC = the parallel system cost per unit time due to downtime or failures
OMC = the parallel system operation and maintenance cost per unit time (i.e., per equipment)

The system downtime for n number of paralleled units is expressed by

$$DT = UT(UA)^n / AV_s \quad (10.27)$$

where

UA = the unavailability of a unit
AV_s = the system availability

Thus, from Equations (10.26) and (10.27) we get

$$TC_P = (OMC)(AV_s)n + (DTC)(UA)^n \quad (10.28)$$

Since $AV_s = 1 - (UA)^n$, from Equation (10.28) we get

$$\begin{aligned} TC_P &= (OMC)[1 - (UA)^n]\, n + DTC(UA)^n \\ &= (OMC)n + [(DTC) - (OMC)n](UA)^n \end{aligned} \quad (10.29)$$

In order to find the minimum value of TC_P, we differentiate Equation (10.29) with respect to n and then equate the resulting expression to zero, i.e.,

$$\frac{d(TC_P)}{dn} = (OMC) + (DTC)(UA)^n \ln UA - (UA)^n(OMC) - (OMC)n(UA)^n$$
$$\ln UA = 0 \quad (10.30)$$

We rewrite Equation (10.30) to the following form:

$$(\text{OMC}) + (\text{UA})^n\,[(\text{DTC})\ln \text{UA} - \text{OMC} - (\text{OMC})\, n \ln n\, \text{UA}] = 0 \tag{10.31}$$

After solving Equation (10.31), we can obtain the value of n for minimum TC_P.

10.9 TIME VALUE OF MONEY, BARRIERS TO OBTAINING USEFUL COST INFORMATION, AND ADVANTAGES AND DISADVANTAGES OF LIFE CYCLE COSTING AND ASSOCIATED IMPORTANT FACTORS

Economics concepts play an instrumental role in life cycle costing because the time value of money has to be considered. What it basically means is that a dollar in hand at present is worth more than a dollar to be received sometime in the future. In life cycle costing, with the exception of acquisition-related costs, the costs occur in the future. Such (future) costs have to be converted to their present values prior to adding them to the acquisition costs. The simplest formula to obtain the present worth of a single future amount is

$$\text{PW} = \text{FA}\left[\frac{1}{(1+i)^m}\right] \tag{10.32}$$

where

PW = the present worth of a sum of money
FA = the future amount of money
i = the compound interest rate per period
m = the number of interest periods (i.e., usually years)

Many other related formulas may be found in various texts on engineering economy [29–32].

10.9.1 EXAMPLE 10.3

Assume that the disposal cost of a military aircraft will be 0.5 million dollars at the end of its 20-year useful life. The estimated value of the compound interest rate is 7% per annum. Calculate the present value of the aircraft disposal cost.

Substituting the specified data into Equation (10.32) yields

$$\text{PW} = (0.5)\left[\frac{1}{(1+0.07)^{20}}\right]$$
$$= \$0.1292 \text{ million}$$

Thus, the present value of the military aircraft disposal cost will be 0.1292 million dollars.

On many occasions, the cost or life cycle cost analysts find it rather difficult to obtain accurate, appropriate, and on time, needed information to perform their tasks effectively. Some of the barriers to obtaining useful cost information within an organization could be as follows [33]:

- poorly designed accounting system architecture
- poorly maintained databases with respect to accuracy
- management overly concerned with secrecy
- highly complex and conflicting reports
- poorly available resources to maintain existing cost-related systems
- inaccurate, late, and poorly formatted reports
- inadequate cost allocation practices
- accounting emphasizing financial rather than managerial information requirements

Just like in the case of any other concept or procedure, life cycle costing also has its benefits and drawbacks. The benefits include comparing competing or alternative projects effectively, selecting competing equipment with a high degree of confidence, reducing the overall cost, and making effective equipment replacement decisions. Similarly, the drawbacks include being costly and time-consuming and giving doubtful data accuracy.

Some of the important factors associated with life cycle costing are as follows:

- Both the equipment manufacturers and users are required to be involved actively in life cycle costing for effective results.
- Good data are required for effective life cycle cost estimates.
- Management is required to play a key role in the practicing of life cycle costing.
- Some surprises may still occur, regardless of the competence of estimators.
- The competence and the experience of the cost estimator could compensate for database-related deficiencies.
- It is absolutely necessary to perform trade-offs between item life cycle cost, performance, and design to cost throughout the project.
- Risk management is the essence of life cycle costing.
- The main goal of life cycle costing studies is to retrieve maximum benefits from meager resources.

- The selected life cycle cost estimation model must have been tailored so that it can include all associated costs in an effective manner.

10.10 PROBLEMS

(1) Discuss the historical developments of the life cycle costing concept.

(2) What are the purposes of performing a life cycle cost analysis?

(3) What are the data required to perform life cycle costing studies?

(4) Outline the steps for conducting a life cycle cost analysis.

(5) What are the skills required to perform life cycle costing studies?

(6) Discuss the elements of a typical life cycle cost estimation model.

(7) Assume that the cost of a 50-megawatt electric generator is $2 million. Estimate the cost of a new similar 90-megawatt generator, if the value of the cost capacity factor is 0.6.

(8) Mean time between failure (MTBF) and mean time to repair (MTTF) of a mechanical pump are 1000 hours and 4 hours, respectively. It is estimated that the pump will be operated for 4500 hours per annum. If the cost of the maintenance labor is $25 per hour, compute the total annual cost for performing corrective maintenance on the pump.

(9) Assume that the retirement cost of an automobile is $700 at the end of its 15-year useful life. If the estimated value of the compound interest rate is 5% per annum, calculate the present value of the automobile retirement cost.

10.11 REFERENCES

1. *Life Cycle Costing in Equipment Procurement*, Report No. LMI Task 4C-5, Logistics Management Institute (LMI), Washington, D.C., April 1965.
2. Dhillon, B.S., *Life Cycle Costing: Techniques, Models and Applications*, Gordon and Breach Science Publisher, New York, 1989.
3. *Life Cycle Costing Procurement Guide (interim)*, Department of Defense Guide No. LCC1, United States Department of Defense, Washington, D.C., July 1970.
4. *Life Cycle Costing in Equipment Procurement—Casebook*, Department of Defense Guide No. LCC-2, United States Department of Defense, Washington, D.C., July 1970.
5. *Life Cycle Costing Guide for System Acquisitions (interim)*, Department of Defense Guide No. LCC-3, United States Department of Defense, Washington, D.C., January 1973.
6. *Acquisition of Major Defense Systems*, Department of Defense Directive No. 5000.1, United States Department of Defense, Washington, D.C., July 1971.

7. Dhillon, B.S., Life Cycle Cost: A Survey, *Microelectronics and Reliability*, Vol. 21, 1981, pp. 495–511.
8. Fabrycky, W.J., Blanchard, B.S., *Life Cycle Cost and Economic Analysis*, Prentice-Hall, Inc., Englewood Cliffs, New Jersey, 1991.
9. Seldon, M.R., *Life Cycle Costing: A Better Method of Government Procurement*, Westview Press, Boulder, Colorado, 1979.
10. Earles, M.E., *Factors, Formulas, and Structures for Life Cycle Costing*, Eddins-Earles, Privately Published, Concord, Massachusetts, 1981.
11. Blanchard, B.S., *Design and Manage to Life Cycle Cost*, Matrix Press, Portland, Oregon, 1978.
12. Dell'isola, A.J., Kirk, S.J., *Life Cycle Costing for Design Professionals*, McGraw-Hill Book Company, New York, 1981.
13. Brown, R.J., Yanuck, R.R., *Life Cycle Costing: A Practical Guide for Energy Managers*, The Fairmont Press, Inc., Atlanta, Georgia, 1980.
14. Dhillon, B.S., *Reliability Engineering in Systems Design and Operation*, Van Nostrand Reinhold Company, New York, 1983, Chapter 9.
15. Reiche, H., Life Cycle Cost, in *Reliability and Maintainability of Electronic Systems*, Edited by J.E. Arsenault and J.A. Roberts, Computer Science Press, Potomac, Maryland, 1980, pp. 3–23.
16. *Life Cycle Cost in Navy Acquisitions*, MIL-STD-259, Department of Defense, Washington, D.C.
17. Dhillon, B.S., Reiche, H., *Reliability and Maintainability Management*, Van Nostrand Reinhold Company, New York, 1985, Chapter 13.
18. *Joint-Design-to-Cost Guide: Life Cycle Cost as a Design Parameter*, DARCOM P700-6 (Army), NAVMAT P5242 (Navy), AFLCP/AFLCP/AFSCP 800-19 (Air Force), Department of Defense, Washington, D.C., 1977.
19. Blanchard, B.S., Fabrycky, W.J., *Systems Engineering and Analysis*, Prentice-Hall, Inc., Englewood Cliffs, New Jersey, 1981, Chapter 17.
20. Kohoutek, H.J., Economics of Reliability, in *Handbook of Reliability Engineering and Management*, Edited by W. Grant Ireson, C.F., Coombs, McGraw-Hill Book Company, New York, 1988.
21. Lamar, W.E., *Technical Evaluation Report on Design to Cost and Life Cycle Cost*, Report No. 165, North Atlantic Treaty Organization Advisory Group for Aerospace Research and Development (AGARD), May 1981. Available from the National Technical Information Service (NTIS), Springfield, Virginia, USA.
22. Coe, C.K., *Life Cycle Costing by State Governments*, Public Administration Review, September/October 1981, pp. 564–569.
23. Bidwell, R.L. *Checklist for Evaluating LCC Program Effectiveness*, Product Engineering Services Office, Department of Defense, Washington, D.C., 1977.
24. Monteith, D., Shaw, B., Improved R, M and LCC for Switching Power Supplies, *Proceedings of the Annual Reliability and Maintainability Symposium*, IEEE, New York, 1979, pp. 262–265.
25. Ganapathy, V., Life Cycle Costing Applied to Motor Selection, *Process Engineering*, July 1983, pp. 51–52.
26. Herd, J.H., Postak, J.N., Russell, W.E., Stewart, K.R., *Software Cost Estimation Study, Vol. 1*, Report No. RADC-TR-77-220, 1977. Doty Associates, Inc., 416 Hungerford Drive, Rockville, MD 20850, USA.

27. Doty, D.L., Nelson, P.J., Stewart, K.R., *Software Cost Estimation Study—Vol. II*, Report No. RADC-TR-77-220, 1977. Doty Associates, Inc., 416 Hungerford Drive, Rockville, MD 20850, USA.
28. Aggarwal, K.K., *Reliability Engineering*, Kluwer Academic Publishers, London, 1993.
29. Thuesen, G.J., Fabrycky, W.J., *Engineering Economy*, Prentice-Hall, Inc., Englewood Cliffs, New Jersey, 1989.
30. Riggs, J. L., *Engineering Economics*, McGraw-Hill Book Company, New York, 1982.
31. White, J.A., Agee, M. H., Case, K.E., *Principles of Engineering Economic Analysis*, John Wiley & Sons, Inc., New York, 1989.
32. Grant, E.L., Ireson, Leavenworth, R.S., *Principles of Engineering Economy*, John Wiley and Sons, Inc., New York, 1990.
33. Richardson, P.R., *Cost Containment: The Ultimate Advantage*, The Free Press, New York, 1988.

CHAPTER 11

Computer-Aided Design and the Information Superhighway

11.1 INTRODUCTION

COMPUTER-AIDED design is concerned with using computers to aid design, and the forecasted amount for the market of computer-aided design, engineering, and manufacturing systems was $25 billion by the middle of the 1990s [1]. Further, it was estimated that by the end of 1983 there were over 44,000 computer-aided design (CAD)/computer-aided manufacturing (CAM) workstations serving the needs of approximately 70,000 engineers and draft personnel [2].

Even though the roots of CAD/CAM may be traced back hundreds of years [for example, Gaspard Monge (1746–1818), a French mathematician, invented orthographic projection in the later part of the eighteenth century], the modern CAD/CAM had its beginnings in 1950 with the demonstration of producing simple pictures using a whirlwind computer at the Massachusetts Institute of Technology (MIT) [3]. Around 1955, numerically controlled (the numerical control approach was pioneered by J. T. Parsons in the 1940s [4]) machines started to take their place in the industrial sector for tasks such as drilling, boring, tapping, and reaming [5, 6]. The late 1950s witnessed the developments of automatic programming for tooling (APT), and one good example of such a system is APT developed by Illinois Institute of Technology for Research (IITRI) [3, 6–8].

The 1960s are regarded as the critical period for interactive computer graphics [2, 9]: the sketchpad system developed at the MIT [10], the GRAPHIC I remote display system's development by Bell Telephone Laboratories [11], the CADAM system's initiation by the Lockheed Aircraft Corporation [3], and the announcement of DAC 1 system (design augmented by computers) by General Motors (GM) in 1964 [3]. Also, it was the decade of the 1960s when the term *computer-aided design*, or CAD, was coined. By the early 1980s, CAD was fully developed in the marketplace, and many people have contributed to its development since its inception [12–14]. Detailed historical reviews of the CAD are given in References [2–4, 6, 15, 16].

Because engineering designers may require various kinds of information from distant sources, the information highway (or superhighway) could be a good means of obtaining needed information. Today, that highway is known as the Internet. The Internet, in turn, is the global network interlinking over 46,000 smaller computer networks and is now over 25 years old and links over 40 million people in 160 countries [17, 18].

The beginnings of the Internet may be attributed to the early 1960s, when computer scientists in the United States started to explore ways and means to directly interlink remote computers and their users [19]. In the latter half of the 1960s, the U.S. government realized the importance of linking researchers in federal laboratories, universities, and some military contractors, and subsequently the U.S. Advanced Research Projects Agency (ARPA) funded a project for this purpose called ARPAnet. The main goal of this project was to develop a network whose communication capability would not be seriously affected in the event of losing the physical sections of the network. As a result, the ARPAnet project led to the development of the Transmission Control Protocol/Internet Protocol, (TCP/IP) software, in other words, the development of language, the network computers use to talk to each other; thus, today's Internet was born. During the 1970s, TCP/IP became the ARPAnet standard network protocol, and the Internet expanded from its four original sites (i.e., servers where information was stored) to 50 sites. Also, during this period the government encouraged educational establishments to take advantage of the ARPAnet system, which resulted in the increase of Internet sites to 200 over the following 10 years.

During the early part of the 1980s, all the existing interlinked Research networks were switched over to the TCP/IP protocol, thus leading to the ARPAnet becoming the backbone of the new Internet. This switching over effort was completed by 1983, and by the February of 1986, the hosts (i.e., computers providing services) linked to the network were increased to 2308. Today, the Internet has millions of hosts connected around the globe [19].

This chapter describes CAD and the information superhighway.

11.2 COMPUTER-AIDED DESIGN

CAD has emerged as an important tool in engineering practice and is being used to fulfill various purposes: producing drawings and documenting designs, generating shaded images and animated displays, conducting engineering analysis on geometric models, and performing process planning and generating numerical control part programs. In particular, with respect to design work, there are various tasks that lend themselves to computer applications, including data manipulation, repetitive calculations, and complex and time-consuming calculations requiring a high degree of accuracy. However, in the design process, the application of the computer depends upon the stage of the design. For example, during early stages such as needs recognition and problem definition, feasibility study, and preliminary design, the use of the computer is rather limited because these stages are more concerned with the creative aspects of the design problem solving. In contrast, the latter stages are better suited for computer applications since they may involve a large number of repetitive and iterative tasks [20].

Over the years in the industrial sector many companies have used CAD technology and reported reduction in product development time and improvement in product quality in addition to many other benefits [21]. Some examples of the benefits accrued by using the CAD/CAM technology in the industrial sector are as follows:

- improvement in product development productivity by 150% [21]
- reduction in development costs by 65% for the IBM 3081 computer [22]
- reduction in the development cycle for new cars from 5 years to 4 or less [23]
- reduction in "Fling 35" camera development costs by 25% [24]
- improvement in productivity (depending on the nature of the task) from 25 to 350% with respect to using CAD systems against drafting boards [25]

11.2.1 CAD/CAM SYSTEM CONFIGURATION AND THE DESIGN PROCESS SUPPORT TOOLS

The five basic elements of a CAD system are graphics, database, program library containing application program, dialog, and data input/output query integrity check. Specifically, the basic elements of any turnkey CAD/CAM or computer graphics system include the following [26]:

- central processing unit (CPU)
- graphics processing unit (GPU)
- graphic workstation(s)
- main memory for CPU/GPU

- hardcopy device (e.g., plotter)
- control unit for the CPU (This may be used in some cases in addition to the workstation.)
- hard disk storage unit (This is used for storing programs and drawings.)
- magnetic tape unit/diskettes (These could be used for system backup, as well as for off loading drawings from mass storage devices.)

There are various kinds of CAD tools required for supporting the design process. For example, a design process composed of six phases will require the tools [3] given in parentheses for each phase: conceptualization (geometric modeling techniques, graphic aids, visualization, and manipulations), modeling and simulation (animation, assemblies, special modeling packages, graphic aids, visualization, manipulations, and geometric modeling techniques), analyses (analysis packages, and customized programs and packages), optimization (structural optimization and customized applications), evaluation (dimensioning, numerical control, tolerances, and bill of materials), and documentation (drafting and detailing and shaded images).

11.2.2 TYPES OF CAD/CAM HARDWARE SYSTEMS

Today, there are several different types of CAD/CAM systems available in the market with varying configurations and options within each type. Since the type of host computer driving the system is the principal factor in this type classification, the CAD/CAM systems are categorized below on the basis of their hardware [3, 27].

- mainframe-based systems
- workstation-based systems
- minicomputer-based systems
- microcomputer-based systems

Once, all CAD/CAM systems, in general, were mainframe-based because it was the only computer type available to drive the systems. One or more design/drafting stations form the mainframe-based CAD/CAM systems. In turn, each design/drafting station, at a minimum, includes a graphics display, a keyboard, and an alpha-numeric control display. Also, a design/drafting station may have input items such as joysticks, lightpens, or other related cursor-control components. In contrast, the mainframe computer is linked to output devices such as printers, plotters, digitizing boards, and storage devices.

A workstation may be described as a "work station" possessing its one computing capability to support important software packages, networking potential with other computing facilities, and multitasking power required by augmented usage and complex and sophisticated tasks [3]. The workstation-based system concept, especially for CAD/CAM applications, offers significant benefits over the timesharing, centralized computing facility accessed through mainframe-based system graphic display terminals. Examples of these benefits are the workstation availability, portability, and time response consistency. It appears that the workstation concept will form the basis of the next generation of CAD/CAM systems.

During the 1970s, minicomputers started to take over from mainframe computers basically because of the cost factor. A typical example of the early versions of minicomputers is the Digital Equipment Corporation PDP series. The rapid growth of the CAD/CAM industry appears to be caused by minicomputers. In comparison to the mainframe-based CAD/CAM systems, the minicomputer-based CAD/CAM systems look similar but with smaller computers. The superminicomputers of today are more suitable for sophisticated CAD/CAM applications than their predecessors because of their speed, accuracy, and storage capacity. Today, a majority of supermini-based CAD/CAM systems available in the market are being sold as turnkey systems (a turnkey system may simply be expressed as a computer hardware and software configuration supplied by a single vendor).

The CAD/CAM field has been significantly impacted by microcomputers, and in particular, the development of the IBM Personal Computer (PC) resulted in major impetus for CAD on PCs. Today, there are various CAD-related software available for use in personnel computers, ranging from two-dimensional drafting to three-dimensional modeling. Two examples of the microcomputer-based CAD/CAM systems are IBM PS/2 and Macintosh II Cx.

11.2.3 CAD/CAM SYSTEM SELECTION AND PROCUREMENT

Because there are many different types of CAD/CAM systems available in the market, a careful consideration is required in procuring them. Even though the needs for acquisition may vary from one buyer to another, the basic selection and procurement process is composed of essentially two primary tasks: preliminary feasibility study and detailed systems analysis and evaluation [28].

The preliminary feasibility study is concerned with determining the appropriateness of procuring design and drafting automation for the organization. The time spent to conduct this study is relatively short. The second task (i.e., detailed systems analysis and evaluation) is the extension and refinement of the preliminary feasibility study and is more comprehensive and detailed in scope.

The main goal of this task is to produce a set of specifications for the acquisition of the CAD/CAM system and to provide a mechanism for evaluating different systems and suppliers.

The overall selection and procurement process is composed of six tasks [28]: conducting preliminary feasibility study, performing information and organization analysis, establishing manual costs definition, establishing systems requirements definition, writing the request for proposal (RFP), and evaluating the system and selecting the vendor. Each of these six tasks is described below.

Preliminary Feasibility Study

The primary objective of this exercise is to review and document the existing operating procedures and practices of the organization and to identify areas most promising to automation, as well as to demonstrate to management the risks and benefits associated with the proposed automation. Some of the issues that must be addressed in this study are growth objectives, in-house technical expertise, expected payback for CAD/CAM, goals/objectives regarding CAD/CAM, potential adaptation of current computer facilities (if applicable); output formats; work task coordination; work task frequency, types, and priorities; differentiation of drafting and other related skill-group tasks; and development of graphic standards (if applicable).

More specifically, the candidate tasks for potential automation should possess characteristics such as critical, labor-intensive, highly repetitive, easy to understand, and definable in terms of input/output procedures. Some of the functions associated with these characteristics are word processing, accounting, project control, and design and drafting leading to automated preparation of engineering design and production drawings.

Information and Organization Analysis

At this stage of the selection and procurement process, the company involved is considering CAD/CAM as a real option. Thus, the goal of this in-depth analysis is to collect information to identify the most profitable CAD/CAM system functions. The analysis should include a further refinement of management's attitudes with respect to the feasibility phase concerning issues such as desired applications for automation, desired payback period, forecasted work flow, purchasing or leasing decisions, and key financial factors. The time required to perform this phase analysis may take somewhere from 3 days to over 6 months, depending upon the conditions peculiar to the organization involved.

Manual Costs Estimation

Just like in the case of any other system procurement decisions, the cost element plays an important role in deciding whether to keep manual or automate the design and drafting-related tasks. The starting point for this task is, first of all, to review the design, drafting, and manufacturing work process diagrams to determine the improvements that can be made. The next step is to cost the manual and proposed automation tasks and then make a cost saving analysis.

The following formula can be used to compute average expected cost for a given task:

$$\mathrm{AEC} = \frac{C_{\max} + 4C_{\mathrm{ib}} + C_{\min}}{6} \tag{11.1}$$

where

AEC = the task average expected cost
$C_{\max}$ = the maximum cost associated with the task
$C_{\min}$ = the minimum cost associated with the task
C_{ib} = the in between or most likely cost associated with the task

System Requirements Definition

After the performance of cost justification analysis, the next step is to decide the CAD/CAM features important to the company. Thus, the main objective of the system requirements task is to establish short- and long range plans to implement CAD/CAM. The short-range plan incorporates objectives for a 1- to 3-year time period, as well as includes information such as estimated cost breakdowns for automation, task description and related information, management priorities, automation associated cost savings, implementation plan, and description of automation effects on management and company. In contrast, the long-range plan is the extension of the short-range plan and includes information on cost and manpower needs for a 3- to 5-year time period.

Since drafting is the common denominator of CAD/CAM systems, the basic factor to be carefully considered is the ease of drawing creation and modification. Some of the documentation, training, and support-related factors are availability and quality of online instruction aids; vendor response time for service calls; quality, cost, and type of hardware and software support; availability of spare parts; and quality and types of documentation provided.

Request for Proposal

RFP allows the organization to choose the most appropriate CAD/CAM system from the available alternatives to meet its multifaceted needs and requirements. The length of an RFP may vary from 15 pages to over 100 pages, depending upon the defining of information requirements by the user organization. The RFP should include enough information so that the vendor/supplier can respond to issues such as general system capabilities, maintenance and support, how and why system capabilities respond to the company's requirements, detailed hardware specifications, warranty conditions, delivery schedules, application software and operating system recommendations and specifications, terms and definitions, installation and environmental needs, acceptance and inspection procedures, and training and documentation recommendations.

System Evaluation and Vendor Selection

There are various important factors relating to software, hardware, and the vendors that are to be considered during the selection process. Some of the important capabilities that are ought to be considered during the CAD/CAM system evaluation and selection include functional capabilities, potential growth, reliability and maintainability, online storage capacity, documentation and user groups, data communications capability, and vendor competence.

With respect to vendor competence, questions should be asked in areas such as

- vendor background
- hardware and software update quality including magnitude and frequency
- cost and quality
- maintenance availability
- flexibility of the vendor with respect to acquiring hardware directly from the manufacturer
- local facility for hardware spare part inventory

The factors above play an important role in the selection of vendor.

11.2.4 CAD/CAM IMPLEMENTATION RELATED SKILL REQUIREMENTS

The introduction of CAD/CAM systems in organizations reduces the labor requirements per unit output but creates the need to upgrade the skills of the involved personnel [29]. These personnel include design engineers, design and drafting technicians, CAD/CAM system development engineers, manufacturing engineers, and manufacturing workers.

The new skill requirement concerning design or other engineers is basically associated with the effective mastery of CAD/CAM tools. The necessity for upgrading the design and drafting technicians' skill is in the areas of abstract problem solving and computer expertise. Even though the needs of engineers in CAD/CAM system development depends to a considerable extent on the make/buy strategy of the organization, the upgrading of the skills among these professionals is necessary as the changes continue in CAD/CAM technology and the organizational requirements.

Probably the most dramatic CAD/CAM related skill upgrading has occurred, because it was necessary, among the manufacturing engineers. Obviously, the main impetus for this upgrading has been manufacturing automation, requiring the necessity to understand and program the new CAM systems. This upgrading caused by CAD/CAM integration resulted in (1) introduction of new complexity into manufacturing automation software and (2) the need of the manufacturing engineers during the integration process to develop a more rigorous producibility constraint characterization, as well as that of the manufacturing planning process.

As the level of the manufacturing automation increases, usually the workers involved require greater job skills. However, the variation often occurs from one type of task to another.

11.2.5 CAD/CAM JUSTIFICATION RELATED MATHEMATICAL ANALYSIS

The introduction of CAD/CAM systems requires a carefully performed economic analysis. Over the years, there have been various methods and approaches developed for this purpose, including the ones discussed below.

Approach I

This is a three-step approach [30] used to make decisions regarding the introduction of CAD in an organization:

(1) Estimate the value of the productivity ratio, using the following relationship:

$$PR = (MH_i - MH_u)/MH_C \quad (11.2)$$

where

PR = the productivity ratio
MH_i = the man-hours of the time prior to the introduction of CAD
MH_u = the man-hours of the time unaffected by CAD
MH_C = the man-hours of the time spent at the console

(2) Estimate the reduction in cost, using the following equation:

$$CR = IBC + (MH_i - MH_u)PC - MH_C(CRC) - MH_C(PC) \quad (11.3)$$

where

PC = the average personnel cost per hour
CRC = the console rate cost per hour
IBC = the indirect benefit cost (under the worst case condition this cost is equal to zero)

(3) To make decisions concerning the introduction of CAD in the organization, use Equations (11.2) and (11.3), along with the factors such as company priorities, past experience, competition, allowable investment, type of organization, and innovation-oriented pressures.

Approach II

This approach was developed by Shah and Yan [31]. The basis for the technique is the benefits and costs associated with the CAD/CAM system operation. The technique can be used in a number of situations: (1) justifying cost for procuring a new CAD facility, (2) estimating the desirable number of CAD workstations for an organization in question, (3) developing the operating strategy for new or already existing facilities, and (4) monitoring the utilization of current facilities. Important assumptions associated with the method are as follows:

- CAD economic analysis is performed in the drawing office environment.
- Increase in productivity of current manpower is the only benefit because the intangibility of the CAD associated benefits (i.e., such benefits cannot be translated into monetary terms).
- Net gain equals benefits minus costs and the net gain is a measure of the economic viability of a given facility.

The benefit D_i expressed in dollars per specified period in producing M_i classification i drawings on CAD versus producing manually is

$$D_i = \mathrm{DC}\,[(\mathrm{SH})\,Qm]\,\theta_i \mathrm{AV}(R_i - 1) \tag{11.4}$$

$$R_i \equiv \mathrm{MH}_i/(\mathrm{CMH}_i/\alpha) \tag{11.5}$$

where

m = the number of work stations associated with the CAD
Q = the number of working shifts per day
SH = the number of working hours in a single shift in the period
DC = the single designer's average man-hour cost
AV = the average system availability factor
θ_i = the fraction (i.e., $0 \leq \theta_i \leq 1$) of CAD system time allotted to classification i drawings
α = the efficiency associated with the designer operating the CAD system
MH_i = the mean number of man-hours required for drafting a classification i drawing, including revisions. This includes, as applicable,extracting data for wiring lists from the drawing through the manual means.
CMH_i = the mean number of man-hours required using the CAD system for drafting a classification i drawing, including revisions. This includes, as applicable, extracting data for wiring lists from the drawing.

Thus, the overall benefit per period is expressed by

$$D = \sum_i D_i \tag{11.6}$$

where D is the total benefit per period expressed in dollars.

Similarly, the total cost, TC in dollars, per period is expressed by

$$\mathrm{TC} = \sum_{j=1}^{6} C_j \tag{11.7}$$

where C_j is the jth cost component; $j = 1$ (user support cost), $j = 2$ (designer training cost with respect to operating the CAD system), $j = 3$ (equipment maintenance cost), $j = 4$ (cost of generating application packages), $j = 5$ (CAD equipment amortized cost), $j = 6$ (CAD installation amortized cost).

If cost, C_{pt}, per terminal-hour of a CAD system is expressed by

$$C_{pt} = \mathrm{TC}/(\mathrm{SH})Q_m \tag{11.8}$$

then the global net gain NG_i in dollars per period if M_i classification i drawings are produced on the CAD system versus manually is given by

$$\mathrm{NG}_i = D_i - C_{pt}/\mu\,(\mathrm{DC}) \tag{11.9}$$

where μ is the system utilization factor and is expressed by

$$\mu = \sum_i \theta_i \tag{11.10}$$

Thus, net gain NG_i is positive only, provided

$$R_i > [C_{pt}/\mu\,(\mathrm{AV})\,(\mathrm{DC})] + 1 \tag{11.11}$$

Since the values of μ and AV are close to unity, the right-hand side of Equation (11.11) simplifies to

$$[C_{pt}/\mathrm{DC}] + 1 \tag{11.12}$$

Thus, we can rewrite Equation (11.11) as follows:

$$R_i > [C_{pt}/\mathrm{DC}] + 1 \tag{11.13}$$

The above relationship can be used to determine the application, if implemented on CAD, that will generate a positive net gain NG_i. The global net gain, NG in dollars, per period is expressed by

$$\mathrm{NG} = \sum_i \mathrm{NG}_i \tag{11.14}$$

It is to be noted that the value of the global net gain will vary at each stage of the CAD implementation phase, depending upon the ongoing activities at that stage.

The mean number of M_i of classification i drawing produced in the time period is

$$M_i - [(\mathrm{SH})(Qm)\,\theta_i \mathrm{AV}(\alpha/\mathrm{CMH}_i)] \tag{11.15}$$

Finally, the number of designers required to produce M_i drawing is expressed by

$$Z_i = Qm\theta_i \mathrm{AV}(\mathrm{SH}/\mathrm{ADH})\left[\frac{\mathrm{ARH}_i}{(\mathrm{CMH}_i/\alpha)} + 1\right] \tag{11.16}$$

where

Z_i = the number of designers required to produce M_i drawing
ADH = the mean number of hours worked per designer in a given time period after considering vacation and sick leave
ARH_i = the mean number of man-hours required for tasks such as planning, preparation, approval, issue, and distribution of classification i drawing produced on the CAD system under study

Approach III

This approach is a ratio basically concerned with estimating the impact of the computer introduction, thus known as the computer impact value. This the ratio, developed by Johnston [31], of the existing weekly cost savings using computers to the total weekly cost savings going from absolutely manual to the optimum level of computer use. In other words, the ratio is a measure of progress and its value is less than unity. The ratio is defined by

$$\mathrm{CR} = (KP_m/P)[\{L(P-1) - 40\,(\mathrm{CC})/(\mathrm{PU})(\mathrm{UT})\}/\{L(P_m-1) - (\mathrm{CC})/3\}] \tag{11.17}$$

where

CR = the computer impact value
UT = the equipment uptime fraction (i.e., $0 \leq \mathrm{UT} \leq 1$)
CC = the computer cost expressed in dollars per people-week
L = the labor rate expressed in dollars per people-week
PU = the mean planned equipment utilization expressed in hours per week. The upper and lower bounds of its value are expressed as $0 \leq \mathrm{PU} \leq 120$.
K = the fraction of work the computers used in comparison to potential uses. The upper and lowerbounds of its value are expressed as $0 \leq K \leq 1$

P_m = the maximum attainable productivity ratio, i.e., rate with the aid of computers in comparison work output to the absolutely manual effort

P = the value of the current productivity ratio

Alternatively, this ratio can also be expressed as

$$\mathrm{CR} = (C_m - C_{pm})/(C_m - C_{\mathrm{cou}}) \tag{11.18}$$

where

C_m = the cost associated with performing all current tasks through manual means

C_{pm} = the cost associated with performing all current tasks in the present manner

C_{cou} = the cost associated with performing all current tasks with optimum computer utilization

This ratio is an excellent tool to provide answers to various management directives and questions: (1) recommend the size of manpower that should be using computers, (2) develop a plan to achieve minimum cost, (3) what is the percent of tasks performed using computers?, (4) make recommendations regarding the percent uptime improvement, and (5) what are the average computer utilization weekly hours?

The two main advantages of this ratio are as follows:

- provides an easy mechanism to comprehend the understanding of the amount of progress achieved in computer utilization
- provides a mechanism to examine cost reduction potential as a quantitative result of changing operational parameters such as productivity, system reliability, and hours worked

Approach IV

This is a simple and straightforward approach of measuring engineering productivity associated with computer-aided design, which was originally proposed by Smith [31]. There are four assumptions associated with the method: (1) engineering hours are the fundamental/basic inputs in terms of effort; (2) engineering drawings released are the principal outputs to be measured; (3) yield is considered in terms of those drawings requiring no changes after their release, however excluding the changes beyond control; (4) equipment availability is measured in terms of the fraction of total scheduled hours available for productive use, but excluding the downtime and training hours.

The following relationship is used to measure progress over the time period:

$$X = A_c . Z_{ad} . D_r \tag{11.19}$$

where

X = the productivity
A_c = the CAD system availability expressed as the ratio of uptime for productive application to total schedule time
Z_{ad} = the acceptable drawing yield expressed as the ratio of acceptable drawings with no revisions to total released drawings
D_r = the drawing release rate expressed as drawings released per hours of engineering and drafting

To make rational decisions, the productivity X against time is plotted.

Approach V

This method is simply composed of a set of indices proposed by Bakey [28] that take design and drafting factors affected by CAD/CAM into account. Each of the proposed indexes is described below.

Index A

This index measures the impact of the CAD system on office space allocation, and the index is defined by

$$\text{SCI} = \frac{\text{ACD}}{\text{ACM}} \tag{11.20}$$

where

SCI = the office space cost impact factor
ACD = the annual CAD system related cost
ACM = the annual manual operation cost

The above index is essentially defined as a ratio between current space cost allocations and the CAD system operating cost under a similar environment. Additionally, it takes into consideration three important factors: (1) space requirement for each operation (i.e., CAD versus manual), (2) the space allocated cost, and (3) the differential between the number of people after installing the CAD system and the current manual operation.

Index B

This index is a measure of effective work output by various people where a single unit or an hour of manual productivity is measured against a fractional unit or fractional hour for the same work done using the CAD system. The index is defined below:

$$\text{PR} = \frac{\text{ML}}{\text{CL}} \tag{11.21}$$

where

PR = the productivity ratio
ML = the manual labor
CL = the labor using the CAD system

The index is useful to determine the additional work an organization can achieve with the current manpower using a CAD system. Using past experience as a reference, the productivity increases of 3:1 to 10:1 is achievable.

Index C

This index is a measure of the susceptibility for the opportunities for error while operating in manual mode as against the CAD system mode. The index is expressed by

$$\text{ERR} = \frac{\text{ME}}{\text{CE}} \tag{11.22}$$

where

ERR = the error reduction ratio
ME = the manual mode potential for error
CE = the potential for error using the CAD system

Index D

This index is a measure of elapsed time prior to fully recovering the CAD system investment amount, that is, through the use of the CAD system. The index is defined by

$$\text{PP} = \frac{\text{CADC}}{\text{CADS}} \tag{11.23}$$

where

PP = the pay out period
CADC = the CAD system cost
CADS = the net monthly savings because of using the CAD system

Four basic elements computed monthly are included in the payout period: (1) cost of the CAD system, (2) maintenance cost of the CAD system, (3) decrease in office space cost, and (4) savings in labor worker-hour cost.

Index E

This index is a measure of total elapsed time for the entire job process taken on a manual basis against the time for the CAD system under study to acquire the same results. The index is defined below:

$$\text{TR} = \frac{\text{MT}}{\text{TCAD}} \tag{11.24}$$

where

TR = the turnaround time ratio
MT = the time taken through the manual process
TCAD = the time taken using the CAD system

In principle, the turnaround time should include the entire process under test or study. Also, before starting the tests, all manual and CAD activities involved in the evaluation process must be clearly identified and documented.

Index F

This index is a measure of difference in operational costs between using the CAD system and the manual means. The index is an effective mean for determining the CAD system impact on the organization's current operating activities. The index is expressed by

$$\text{DCI} = \frac{\text{YCADC}}{\text{YMC}} \tag{11.25}$$

where

OCI = the operational cost impact ratio
YCADC = the yearly operating cost associated with the CAD system
YMC = the yearly operating cost associated with the manual process

The operational cost impact includes the computation of various diverse components: (1) maintenance cost of the CAD system, (2) cost of the CAD system, (3) savings in the labor cost by using the CAD system, and (4) costs associated with the CAD system installation and manpower training.

11.3 INFORMATION SUPERHIGHWAY

The rapid advances in electronic communication technologies have paved the way for the development of the information superhighway. In fact, the term *information highway*, or *superhighway*, is rapidly entering our vocabulary; for example, according to Reference [18] in January 1993, the term *information highway* or its variations appeared only 57 times in articles stored in the Nexis database (information in this database is collected from hundreds of magazines, newspapers, and other related sources), but a year later in January 1994, its usage skyrocketed to 1480 times. Although the term *highway*, or *superhighway*, is increasingly being used in our day-to-day vocabulary, there is no, as yet, universally accepted definition of it.

Vice President Al Gore originally proposed the information superhighway scheme in 1979, and more than a decade later in December 1993 stated the involvement of four major groups with respect to a new information marketplace based on information highways: information highway owners, information customers, information providers, and information appliance manufacturers [32].

The information superhighway is anticipated to have a wide range of applications and will affect public institutions, the private sector, and private citizens. Public institutions are expected to use the information highway to improve the delivery of public services and allow citizens to have easy access to government information. The usage of the information superhighway by the private sector could lead to the formation of more effective teams, with members located at various places, thus allowing flexibility for businesses to be located virtually anywhere. Also, the superhighway will allow an opportunity to set up various value-added, information-related services. Private citizens will use the information superhighway to gain access to up-to-date information through news, on-line databases, at-home shopping, video on demand, and more.

According to Reference [32], each element of the information superhighway should follow principles such as promoting and encouraging competition, promoting open access to networks, encouraging private investment, promoting flexibility, and avoiding the creation of a society of information "haves and have nots."

Vice President Al Gore [33] predicted that the information superhighway will turn out to be the most important and lucrative marketplace of the next century. This message is further emphasized by John Sculley, Apple Computer chairman, who stated that this mega-industry could reach $3.5 trillion worldwide by the year 2001. In the future, the information superhighway is expected to play an increasing role in the area of engineering design. The current Internet may probably be called an initial form of the information superhighway.

11.3.1 INTERNET

Today, the Internet has become world's biggest computer bulletin board and data bank, linking millions of people who use it for various purposes, including sharing research results, sending and receiving mail, playing games with opponents in other cities, sharing gossip, and searching for information in hard-to-reach library facilities [33].

In the usual sense of the word, the Internet is not owned by anyone; however, its backbone in the United States is the National Science Foundation (NSF), which provides funds. The technical support is provided by the Internet Engineering Task Force (IETF), a committee of scientists and experts. In addition, there are many regional and international components of the network that have their own administration and funding. However, there is an Internet Architecture Board (IAB) that sets standards and makes decisions concerning the Internet that have to be abided by any network connected to the Internet.

In 1993, three 5-year cooperative agreements for managing the Network Information Services were signed by the NSF. More specifically, the recipients of these three agreements agreed to manage the Internet Network Information Center (Inter NIC) jointly, in other words, providing information concerning getting connected to and using the Internet. The three recipients were Network Solutions, AT&T, and General Atomics [19]. Network Solutions agreed to provide the Internet registration services and AT&T to maintain lists of file transfer protocol (FTP) sites, a list of servers available on the Internet, library catalogs, data archieves, database design, management, maintenance services, and so on. General Atomics was assigned to provide general information about the Internet and network educational services.

Over the years, a number of organizations has been established with membership involved in educating people regarding the Internet or exploring subjects vital to the Internet [19]:

- Internet Society (ISOC): This is a nonprofit organization and is the secretariat of the IETF and IAB. The Internet Society holds its meetings annually, which includes symposia and workshops on the topics of current interest.
- Corporation for National Research Initiatives (CNRI): This is a nonprofit organization that encourages cooperation among government, academic, and private sector people in the development of a national data network. CNRI is involved in organizing research projects such as faster transmission lines for carrying live video broadcasts and graphic simulations, knowbots (i.e., programs similar to good computer viruses and powerful to search through the Internet for the required information by the knowbots "owner"), and more.
- Electronic Frontier Foundation (EFF): This organization is concerned with the social effects of computers on society, in particular, the legal rights of computer users. EFF is seriously involved in shaping public policy concerning computer-based communications.
- Computer Professionals for Social Responsibility (CPSR): This organization was formed in 1983 and is basically concerned with the ethical use of computers, including issues such as the privacy of personal online information, the reliability of software controlling possibly life-threatening processes (e.g., nuclear reactors), pollution generated by the manufacture and use of computers, and the replacement of workers by computers.

New Developments

In further development of the Internet, the U.S. government is playing a key role. For example, it has committed $400 million to build the National Research and Education Network (NREN) that will start by making use of the existing U.S. Internet. During the first phase, it is expected that the NREN will connect over a dozen leading research centers through a 1- to 3-gigabit-per-second fiber-optic-based network using basically the current fiber-optic cables [34]. In the final phase, it is expected that NREN will displace the existing Internet, which is rather slow to transmit large texts such as databases, complete books, high resolution graphics, and so on.

Justifications, Benefits, and Difficulties

In deciding for an organization to have a direct link to the Internet, there are a number of factors that must be considered [18]:

- perspective: Management is generally inclined towards strategic solutions to strategic problems. Thus, in order to win the backing of the management approach, the Internet issue is not from a technology perspective, but from a strategic perspective.
- opportunities: A clear understanding of the Internet opportunities for the organization is necessary. Also, the knowledge concerning the standing of similar or competitive organizations on the Internet issue is useful to examine one's own position.
- expectations: These must be considered from realistic point of views. Instant riches or revolutionary results may not be possible through the introduction of the Internet in an organization.
- capability: This is concerned with determining the "inter-network" capability of the organization or determining the necessary steps required to become capable.
- limitations: The Internet is one technology tool to help the organization. It is useful to clearly understand the role of all communication technologies within the organization from a strategic and business perspective rather than just from the Internet.

The use of the Internet is growing at an alarming rate throughout the world as many organizations experience benefits such as a reduction in time to do business globally, an efficient customer service tool, a common electronic mail link between companies, an invaluable global marketing tool, and reduction in communication costs globally.

Internet may not be a complete solution to all information needs of an organization or an individual. Some of the shortcomings associated with the Internet are as follows [18]:

- relevance of the available information: Since the Internet had its roots in academic and research communities, a large amount of the available information on the Internet tends to be from these communities. Thus, a general user may find it quite frustrating to obtain the needed information
- nonpermanent nature of many Internet information resources
- sometimes questionable reliability of the information obtained using the Internet

- difficulty locating the required information because of nonexistence of a "master index"
- difficulty obtaining in-depth information on certain subjects

The Legal Issues and Some Related Guidelines for Engineers

The rapid growth in the use of the Internet has given rise to various legal issues that directly or indirectly concern engineers. Some of the items directly or indirectly concerning the Internet and engineers are discussed below [35]:

- software piracy: The Internet is a haven for software piracy; for example, in 1994, the cost of piracy to American and Canadian software developers was on the order of $2 billion (i.e., including on the Internet). The Internet allows electronic bulletin boards and host computers to store stolen software at no cost, for distribution worldwide. Recently, for example, a Massachusetts Institute of Technology (MIT) engineering student set up an electronic bulletin board on the Internet to encourage software pirates to store popular stolen software so that others can copy it. As a result of his action, it is estimated that over $1 million in licensing fees and royalties were lost.
- copyright and trademark infringement: Under the current law, copyrighted material cannot be copied onto the Internet without having prior permission of the owner. Thus, it is important for engineers contemplating going online to take appropriate measures against unlawful company trademark usage. Recently, there were many cases of copyright and trademark infringement in various areas involving the Internet: music (*Frank Music Corp. versus CompuServe Inc.*), videos (*Sega Enterprises Ltd. versus Maphia et al.*), photographs (*Playboy Enterprises versus Event Horizon*), and trademark rights (*Playboy Enterprises versus Ferna*).
- confidential information: There is no privacy of information stored/transmitted on the Internet because it can be intercepted, copied, and altered. As the U.S. and Canadian governments treat some encryption technologies as being sensitive or secret, the placing of encryption softwares on the Internet could fall in the category of prohibited export, thus leading to prosecution under export control laws.

Since the corporate records, financial information, employee records, and intellectual property are susceptible to unlawful use, tampering and viral contamination, and corporate espionage on the Internet, engineers involved can take measures to minimize these threats [35]:

- Eliminate the transmission of confidential or proprietary business information through E-mail.

- Make use of electronic "firewalls." These are dedicated computers and software used for the purpose of separating susceptible computer networks from Internet access.
- Protect the intellectual property owned by the employers by acquiring and registering intellectual property rights.
- Develop employee screening and training programs with respect to protecting access codes and confidential and proprietary business information.
- Develop a mechanism for involved employees to sign nondisclosure agreements on competitive and confidential information.

11.3.2 COMMON INTERNET SERVICES

There are many commonly used Internet services. In fact, the commonly used services may be grouped as follows [19]:

- information search: This category includes services such as wide area information server (WAIS), Archie, and Veronica.
- communication: This category includes services such as electronic (E) mail, Telnet, Users' network (UseNet), and Internet relay chat (IRC).
- information retrieval: This category includes services such as Gopher and File transfer protocol (FTP).
- multimedia information: This category includes the World Wide Web (WWW) service.

Each of the above services is discussed below [18–20, 36, 37].

Wide Area Information Server

WAIS is pronounced as "ways" and is a system that searches the subject under study through documents on servers located all around the world. More clearly, the WAIS system searches databases indexed with keywords and then lists addresses for locating documents of interest. The indexes to all WAIS servers on the Internet are maintained by a central site (i.e., on the Internet) that can be used as the search starting point.

Archie

This was one of the first Internet information retrieval systems and was developed by the Computing Centre, McGill University, Canada. Archie creates a central index of files available on anonymous FTP sites throughout the Internet system. It is to be noted that there are over 800 anonymous Internet FTP Archie sites containing around 1 million files representing over 100 gigabytes of information. Thus, Archie is a useful tool to find a specific file on the Internet.

Once a month, the Archie server updates the listing information from each site, thus providing users information reasonably in a timely manner.

Veronica

This is a gopher service analogous to Archie for anonymous FTP sites with a key difference that the Veronica system is specifically designed to locate keywords at the gopher sites established on the Internet. It is usually easier to find the desirable information using Veronica than Archie because gopher menu items may be descriptive phrases rather than just file names.

After the completion of searching the gopherspace, Veronica creates a gopher menu of all the items found, which can be used to select desirable items. To use Veronica, a connection to a gopher server is required, because it is the gopher server that provides access to a Veronica server.

Electronic (E) Mail

E-mail is the most pervasive and popular Internet application and one of the first Internet services developed. It allows for the composition and transmission of messages electronically between people on either local or remote machines. Some of the Internet services may be accessed by sending an E-mail request and delivered as an E-mail message.

Telnet

This is one of the early Internet access tools. Telnet allows an interlink with a computer at a remote location to read that computer's directory of files, as well as to use its programs. In other words, an authorized user using the Telnet tool can log on to a remotely located computer and have much the same access to it as to his/her own computer. Telnet has proven to be a valuable scientific and engineering research and development tool that permits individuals around the world to work together and share resources.

Users' Network

Usenet was developed at Duke University and the University of North Carolina and allows the discussion of subjects of mutual interest with Internet users located anywhere around the world. Basically, the Usenet sites store and transmit news items to other Usenet sites located on the Internet. For example, when a news item is posted on the local computer, first it is stored on the computer and then sent to other computers that agree to exchange netnews items with the local computer. In turn, these computers forward the article to other computers, who send it to others, and the process continues until the news item reaches all the participating Usenet computers.

Internet Relay Chat

IRC was developed in the late 1980s with the aim of replacing the UNIX talk program, and the service allows multiple people to communicate simultaneously. IRC is a client/server application similar to various other internet services. An IRC client is required to be activated when the need arises to communicate among individuals, in addition to getting connected to an IRC server. Once on the server, a desirable channel is selected. This is a good means of having a real-time conference and allowing every one to participate equally, but the communication speed is quite slow.

Gopher

Gopher is an extremely popular Internet tool that may be described as one of the main "surfboards" for the Internet. The Gopher protocol and software package was developed by the University of Minnesota, and the name itself was derived from the University's school mascot. According to one study [18], the population of Gopher servers on the Internet is multiplying at an alarming annual rate of 997%. This dramatic increase may be due to the fact that the Gopher allows you to retrieve data without having a clear-cut objective. More clearly, the basis for the Gopher navigation is to enquire about a subject, and the navigation itself does not depend on having computer addresses. Obviously, these characteristics make the Gopher an extremely useful introductory tool for new Internet users. Specifically, the Gopher is a useful service to locate information on Internet resources, complete guides to the Internet, specific mailing lists, documents describing the use of various Internet services, and locations of software within particular FTP sites.

Additional, attractive characteristics of most Gopher systems are the capability to download the document to a local computer (i.e., information seeker's computer), transfer the file to the Internet service provider for downloading it later, and have the viewing document sent to the viewer through E-mail.

File Transfer Protocol

FTP is one of the first developed Internet services that permits individuals to transfer files between computers, usually connected through the Internet. It means that a computer system having FTP and connected to the Internet can access various types of files available on other computer systems. Examples of the files that could be transferred are application programs, text, software updates, binary codes, pictures, and sound. Many computer systems connected to the Internet make files available through anonymous FTP, which means those machines could be accessed without having accounts on them.

World Wide Web

WWW, or Web, is one of the most recent Internet services that was developed at the European Laboratory for Particle Physics (CERN). In 1989, some scientists at CERN expressed the need to have a better access mechanism to shared information by widely dispersed research groups. As the result of this need, the Web was released for general use at CERN, and in 1992, CERN started to publicize its accomplishment.

The Web permits combining text, graphics, audio, and animation to produce a document, and the links within Web documents can take the information seeker promptly to other related documents. Since 1993, several browsers had been developed for various computer systems: Apple Macintosh, Personal Computer (PC)/Windows, and X Windows. A browser may simply be described as a software application capable of interpreting and displaying documents that it finds on the Web. Some of the Web browsers are Mosaic, Lynx, and Netscape.

11.4 PROBLEMS

(1) Write an essay on the history of CAD.
(2) Describe the basic elements of a CAD system.
(3) Discuss the four types of CAD/CAM hardware systems.
(4) Describe the CAD/CAM system selection and procurement process.
(5) Cite at least four examples of the benefits experienced in the industrial sector when using the CAD/CAM technology.
(6) Write an essay on the historical developments of the Internet.
(7) Discuss the benefits and the shortcomings associated with using the Internet.
(8) What are the important legal issues associated with Internet?
(9) Describe the information superhighway.
(10) Discuss at least three of the most important Internet services.

11.5 REFERENCES

1. Robertson, D.C., Managing CAD Systems in Mechanical Design Engineering, *IEEE Transactions on Engineering Management*, Vol. 39, 1992, pp. 22–31.
2. Machover, C., CAD/CAM: Where It Was, Where It Is, and Where It Is Going, in *CAD/CAM Handbook*, edited by Teicholz, E., McGraw-Hill Book Company, New York, 1985, pp. 2.1–2.27.
3. Zeid, I., *CAD/CAM Theory and Practice*, McGraw-Hill, Inc., New York, 1991.
4. Besant, C.B., Lui, C.W.K., *Computer-Aided Design and Manufacture*, John Wiley & Sons, New York, 1986.
5. Pease, W., An Automatic Machine Tool, *Scientific American*, 1952, pp. 101–115.
6. Bezier, P., First Steps of CAD, *Computer-Aided Design*, Vol. 21, 1989, pp. 259–261.
7. Brown, S., Drayton, C., Mittman, B., A Description of the APT-Language, *CACM* 6, 1963, pp. 649–658.
8. Coons, S., An Outline of the Requirements for a Computer-Aided Design System, *AFIPS (SJCC)*, Vol. 23, 1963, pp. 299–304.

9. Chasen, S.H., Historical Highlights of Interactive Computer Graphics, *Mechanical Engineering*, Vol. 103, No. 11, 1981, pp. 110–115.
10. Sutherland, I., SKETCHPAD: A Man-Machine Graphical Communication System, *AFIPS (SJCC)*, Vol. 23, 1963, pp. 329–346.
11. Ninke, W., GRAPHIC 1: A Remote Graphical Display Console System, *AFIPS (SJCC)*, Vol. 22, 1965, pp. 839–846.
12. Elliott, W.S., Computer-Aided Mechanical Engineering: 1958–1988, *Computer-Aided Design*, Vol. 21, 1989, pp. 275–288.
13. Eastman, C.M., Architectural CAD: A Ten Year Assessment of the State of the Art, *Computer-Aided Design*, Vol. 21, 1989, pp. 289–292.
14. Robertson, D.C., Managing CAD Systems in Mechanical Design Engineering, *IEEE Transactions on Engineering Management*, Vol. 39, 1992, pp. 22–31.
15. Llewelyn, A.I., Review of CAD/CAM, *Computer-Aided Design*, Vol. 21, 1989, pp. 297–302.
16. Encarnacao, J., Schlechtendahl, E.G., *Computer-Aided Design*, Springer-Verlag, Berlin, 1983.
17. Di Stefano, B., One Engineer's Journey on the Information Highway, *Engineering Dimensions*, Vol. 16, No. 6, 1995, pp. 24–27.
18. Caroll, J.A., Broadhead, R., *Canadian Internet Handbook*, Prentice-Hall Canada Ltd., Toronto, 1994.
19. Pike, M.A., et al., *Using the Internet*, Que Corporation, Indianapolis, 1995.
20. Dhillon, B.S., *Engineering Design*, Richard D. Irwin, Inc., Chicago, 1996.
21. Robertson, D.C., Managing CAD Systems in Mechanical Design Engineering, *IEEE Transactions on Engineering Management*, Vol. 39, 1992, pp. 22–31.
22. Swerling, S., Computer-Aided Engineering, *IEEE Spectrum*, Vol. 19, 1982, pp. 37–41.
23. Fitzgerald, K., Compressing the Design Cycle, *IEEE Spectrum*, Vol. 24, 1987, pp. 39–42.
24. "The Fling 35 Story—Success with CAD/CAM", Kodak Corporation Video Productions, Eastman Kodak Co., U.S.A., 1989.
25. Machover, C., Blauth, R.E., *The CAD/CAM Handbook*, Published by Computervision, Bedford, Massachusetts, 1980.
26. Stover, R.N., *An Analysis of CAD/CAM Applications*, Prentice-Hall Inc., Englewood Cliffs, New Jersey, 1984.
27. *CAD/CAM Market in the United States*, Report No. 1564, Frost and Sullivan, Inc., New York, 1986.
28. Teicholz, E., System Selection and Acquisition, in *CAD/CAM Handbook*, edited by Teicholz, E., McGraw-Hill Book Company, New York, 1985, pp. 6.3–6.20.
29. Adler, P.S., CAD/CAM: Managerial Challenges and Research Issues, *IEEE Transactions on Engineering Management*, Vol. 36, 1989, pp. 202–215.
30. Chasen, S.H., Dow, J., *The Guide for the Evaluation and Implementation of CAD Systems*, CAD Decisions, Atlanta, Georgia, 1979.
31. Preston, E.J., Crawford, G.W., Coticchia, M.E., *CAD/CAM Systems*, Marcel Dekker, Inc., New York, 1984.
32. Brassard, D., *Information Superhighway*, Report No. BP-385E, March 1994, Science and Technology Division, Research Branch, Library of Parliament, Ottawa, Canada.

33. Ripley, C., editor, *The Media and the Public*, The H.W. Wilson Company, New York, 1994.
34. Powell, D., Supernetworks in Canada Play Catch-Up, *Computing Canada*, February 1992, p. 6.
35. Thomson, G., Engineers and the Internet: The Legal Issues, *Engineering Dimensions*, Vol. 16, November/December 1995, p. 32–33.
36. Learning the Language, *Engineering Dimensions*, Vol. 16, November/December 1995, p. 28.
37. Maxwell, C., Grycz, C.J., *Internet Directory*, New Riders Publishing, Indianapolis, 1994.

Index

Biography

DR. B.S. DHILLON is a full Professor of Mechanical Engineering at the University of Ottawa. He has also served for over a decade as Chairman of the Mechanical Engineering Department or Director of the Engineering Management Program at the same institution. He has published over 250 articles on reliability engineering and related areas. He is or has been on the editorial boards of several journals. In addition, he has written 17 books on various aspects of system reliability, safety, human factors, maintainability, and engineering management, published by Wiley (1981), Van Nostrand (1982), Butterworth (1983), Marcel Dekker (1984), Pergamon (1986), and more. His books on reliability have been translated into several languages, including Russian, Chinese, and German. He served as General Chairman of two international conferences on reliability and quality control, which were held in Los Angeles and Paris in 1987.

Dr. Dhillon is a recipient of the American Society for Quality Control Austin J. Bonis Reliability Award, the Society of Reliability Engineers' Merit Award, the Gold Medal of Honor (American Biographical Institute), and Faculty of Engineering Galinski Award for Excellence in Reliability Engineering Research. He is a Registered Professional Engineer in Ontario and is listed in American Men and Women of Science, Men of Achievements, International Dictionary of Biography, Who's Who in International Intellectuals, and Who's Who in Technology.

Dr. Dhillon has been teaching engineering design for many years and attended the University of Wales, where he received a B.S. in electrical and electronic engineering and an M.S. in mechanical engineering. He received a Ph.D. in industrial engineering from the University of Windsor.